Battles

and

Battleships

a narrative history of

warships and naval

warfare

from 1866 to 1905

written by Peter Stetson

co-authored and edited by Paul Janson

Those who do not learn history are doomed to repeat it

*George
Santayana*

*Those who do learn history will still repeat it, but
they will know they are repeating it*

*Paul
Janson*

*Men make their own history, but they do not
make it as they please; they do not make it under self-
selected circumstances, but under circumstances
existing already, given and transmitted from the past.*

*From the Eighteenth Brumaire of Louis
Bonaparte, 1852*

*Karl
Marx*

Battles and Battleships

Published by JM Publishing

36 Elm Street, Georgetown, MA 01833

Front Cover: CSS *Virginia* and the sinking of the USS *Cumberland*, March 8, 1862. The first ironclad to engage in naval battle at sea and to sink a seagoing enemy vessel.

The CSS *Manassas* was an ironclad riverboat. She was launched in 1855 in Medford, MA as the *Enoch Trainer*, a steam tug/ice breaker that was converted to a ram by the Confederate States of America in 1861 with the addition of 1¼-inch iron plate armor and a single bow-firing gun. She engaged US naval vessels on the Mississippi River on October 12 of that year, ramming the USS *Richmond* making her the first ironclad to engage in battle, not a sea battle, but a river battle, leaving room for her navy's ship *Virginia* to claim that honor.

I lived in Medford for a while and this minor first was never mentioned there. Amelia Earhart once (briefly) lived in the city a plaque announces, and they were very proud of the assertion that the popular wintertime song, "Jingle Bells", was probably composed there, although James Lord Pierpont, the composer, waited until he had moved to snowless Savannah, GA to publish and copyright it in 1857. It was, incidentally, considered quite risqué at the time.

Savannah deserves mention in maritime annals as she gave her name to the first steam powered ship, a merchant vessel, to cross the Atlantic Ocean (1819), and the first nuclear powered merchant ship (1959). Neither was commercially successful, unlike Mr. Pierpont's song, which certainly was and still is. The NS (Nuclear Ship) *Savannah* may be seen at Pier 13, Canton Marine Terminal, Baltimore, Maryland.

Back Cover: The USS Missouri fires a test salvo during World War II. She was one of last battleships built for the US Navy, and is now a museum ship at Pearl Harbor, HI, ironic since Hawaii is one of two US states to never have a battleship named for her - Alaska is the other.

The shells being fired by the *Missouri* on the back cover cost about $100,000.00 each or $900,000.00 total for this nine-gun salvo in 1943. Comparing this to the approximate $300,000.00 total cost for the construction of the USS *Constitution* (adjusted to $420,000.00 in 1943 dollars by using the adjusted CPI method), would illustrate that this single test salvo would have paid for the construction of two *Constitution* sized warships. The tomahawk missile, placed on Iowa class battleships during the 1980's, cost $1.41 million each, or $12.7 million for a salvo of nine, which is incidentally the estimated cost of the repairs now underway on the USS *Constitution*. .The cost of war has changed over the years.

**Cover design by Elm Street Design Studio,
Rick Schroeppel www.elmstreetdesignstudio.net**

Copyright Paul Janson July 2016, JM Publishing

Acknowledgment

This book was researched and written by my cousin, Peter Stetson. I have made editorial adjustments where they seemed to be necessary, but those were few and I found no cause to tamper at all with the body of this work. Unfortunately, Peter died shortly after this manuscript was completed, otherwise he might well have done what I have done by way of editing it himself. He spoke to me of his gratitude to the library staff at the Wakefield Public Library in Wakefield, Massachusetts and so I will pass this along to them here as well as the staff at the Georgetown Peabody Library in Georgetown, Massachusetts for their invaluable assistance to me.

Paul Janson

Table of Contents

PROLOGUE

It was the morning of March 9, 1862. A battle was taking place that would change naval warships forever. Two ships were engaged in Hampton Roads on the James River in the state of Virginia, in the then Confederate States of America. Although the battle itself did little more than continue the status quo of the American Civil War, few if any encounters in warfare have brought about so great a change in a single day.

The two ships that faced each other were the CSS *Virginia* and the USS *Monitor*. The *Virginia* is also commonly referred to as the *Merrimack*, the name of the steam frigate that was burned and scuttled by the retreating US Navy during the opening moves of the war to prevent its capture. It was however, raised by the Confederate Navy and fitted with iron plates making her an "ironclad". The USS *Monitor* was more innovative although hastily built (only 101 days) in response to the rumors that the *Virginia* was being built. One of the more remarkable things about this encounter is that the ships were completed and present for action within 24 hours of each other.

The idea of an ironclad warship was not new in 1862. The French had built the ironclad *Gloire* in 1859 and the British had launched HMS *Warrior* in 1860. *Warrior* was an iron-hulled armored ship and her obvious advantages prompted the British Navy to decide that it should have an all-armored fleet in 1861. The impetus for this change was the wide use of explosive shells capable of not only penetrating the wooden hulls of existing ships-of-the-line, but setting those wooden ships afire. During the recently fought Crimean War, the Russian Navy had demolished an Ottoman fleet of wooden vessels with explosive shellfire at the Battle of Sinop on Nov 30, 1853.

The Confederate States had actually sent the CSS *Manassas,* an ironclad riverboat/ram, into battle against the US naval vessels on the Mississippi River on October 12, 1861 where she rammed, but failed to sink the USS *Richmond*. *Manassas* was therefore the first ironclad to engage in battle, albeit not a sea battle, The battle at Hampton Roads that March morning took that honor with *Virginia* successfully engaging ocean going warships and the results of that engagement left no doubt about the future of naval development.

The ships involved were as unlike each other as might have been possible. The *Virginia* was essentially a wooden ship with 1 - 4 inch iron plate bolted to its sides. Its guns were two 7-inch Brooke rifles, two 6.4-inch Brooke rifles, six 9-inch Dahlgren smoothbores and two 2 12-pounder howitzers. She was also fitted with a ram. Her maximum speed was 5-6 knots and her draft was 21 feet. Her deep draft would limit her threat since she could not proceed up the Potomac River to bombard Washington DC as was feared at the time. It would also be her ultimate undoing since as the Federal Army moved up the James River toward Richmond; the Confederate Navy was forced to scuttle her (once again) to prevent her capture.

The *Monitor* in contrast was the work of the brilliant naval engineer John Ericsson and contained a number of innovations not the least of which was the rotating turret. She was an iron built ship with a speed of 6 knots. Her armament was two 11-inch smoothbore Dahlgren guns, armor: $1 - 9$ inches (8 inches on the gun turret).

The battle between these two lasted four hours and showed no clear advantage to either vessel. They were both pounded at close range with no significant result and a stalemate followed until the *Virginia* was scuttled on May 11 by the retreating Confederates. Perhaps more significant than this first encounter of ironclads, which established that current guns were unable to seriously damage armored vessels, was the action that took place on the day before.

The *Virginia* first descended into Hampton Roads on March 8th to engage the blockading federal fleet. In that brief encounter, the sail frigates *Cumberland* and *Congress* were sunk and the steam frigate *Minnesota* was run aground to avoid being sunk. The *Virginia* was returning the next morning to complete the destruction of the blockading fleet when she encountered the *Monitor*. The action on the eighth of March showed unequivocally that wooden ships could not stand against armored vessels. This made every navy in the world obsolete, and future naval design and warfare was dictated accordingly.

A few additional observations are worth noting here. First neither the *Monitor* nor the *Virginia* were heavily armored by standards common in a few years, but neither was threatened by the armament they faced. Second, while they engaged each other on an inlet of the Atlantic Ocean, neither vessel was seaworthy. Their respective navies never attempted to sail them upon the ocean, *Virginia* never left the James River and the *Monitor* was towed to Hampton Roads and sank while being towed from it. Yet for all of this, they are arguably the most famous warships of all time, recognized by name by the most casual of history students.

A note of interest regarding the battle, it lasted about four hours and, as already said, no significant damage was inflicted on either vessel. The *Virginia* fired explosive shells, while the *Monitor* fired only solid shot. It took about eight minutes to reload and fire the guns even though they engaged each other at essentially point blank range. The gunners in Admiral Nelson's command at Trafalgar, 60 years earlier, would have been embarrassed by such a dismal rate of fire.

This volume hopes to chart the course of naval development through the period from the Battle of Lissa in 1866 to the Battle of Tsushima in 1905. Those forty years saw the development of battleship design to its final form, the HMS *Dreadnought*. The evolution was in many cases haphazard, dictated by advances in technology, particularly metallurgy, and often by the idiosyncratic nature of the designers and purchasers of warships. The final proving ground of each innovation was the naval battle and the future was determined by the results and by the ability of observers to interpret the results. Thus, the battles are always the central point of examination.

The lessons were not always clear and not uniformly adopted, not without strong dissenting voices at least. This is the account of that development and I hope that the parallels to military development in the past and present will be clearly recognized. The serious student will easily take a lesson from this comparison.

Finally, this is a narrative history, detailed in nature and researched in depth, but is not intended to be a true reference source. Rather it is intended to give an essential history and the sometimes fascinating details with direction to the reference sources used to compile it. I hope this will read easily and that the bibliography will provide a starting place for further research by those who wish to pursue that avenue in more detail.

GENERAL INTRODUCTION

In this section, I hope to briefly establish the terms used and clarify their context in order to make this reading both more understandable and more enjoyable. The context will be preserved as much as clarity allows and that will add to the understanding and enjoyment. To say that: "… the two men repaired to their respective commands to prepare to receive the enemy." is not the same as saying: "… they returned to their ships to wait for the enemy." The latter unnecessarily, but successfully, avoids the drama that is deserved. Likewise, the standard language of the time and place is used here for similar reasons, and an explanation is provided only where the context fails to allow for an accurate interpretation. A vessel that is damaged and "is shipping considerable water" is clearly leaking to the verge of flooding. No explanation will add to the text.

First, an understanding of the outside forces that directed the development of warfare in general and warships in particular, will contribute considerably to the appreciation of the discussion. Advances in technology, particularly metallurgy, were the major contributors to the development of warships. These technological advances dominated the design of the ships, their armor and the guns they possessed, and nearly every other aspect of their construction.

A brief discussion of the stuff of which the vessels themselves were made is a good place to begin. The standard naval vessel in the era before this book begins was a wooden ship powered by sail. If it were to earn the designation of "ship-of-the-line", that is if it were to be suitable to stand in the line of battle, it would mount between 64 and 120 smoothbore guns almost exclusively along its sides. These were generally fired at closed range, 50 yards, and in a volley termed appropriately: a broadside. They fired a number of different loads, but the most common was a solid shot: a cast iron ball. The weight of this ball designated the size of the gun and these were between 6 and 32 pounds. It was necessary for these weapons to be inside the ship to load them, and therefore they were relatively short-barreled, and this limited the effective range and accuracy. Once loaded they were rolled out through ports and fired while protruding from the ship's hull. This improved the field of fire and minimized blast damage from the discharge. The recoil would assist in bringing the gun back inside to be reloaded.

The largest of these guns, the 32-pounders, were mounted on the lowest deck, only a few feet above the waterline. This was necessary because of their weight, almost 8000 pounds, and their recoil. Mounting these guns on the upper decks would have subjected the ship to too much stress and made her so top-heavy as to risk capsizing in all but the most tranquil of seas. The crews under the command of Admiral Nelson at Trafalgar in 1805 could load and fire these weapons every ninety seconds. This was faster by a margin of fifty percent compared to the French and Spanish opposing him and that was a significant advantage for the British. HMS *Victory*, Nelson's flagship, mounted 104 guns and has been preserved in Portsmouth, Hampshire, United Kingdom where she may be toured. By way of comparison USS *Constitution* was a frigate, rated at 44 guns, but mounting as many as 52 guns, still not sufficient to be classed as a ship-of-the-line. She may be toured in Boston, Massachusetts.

The nature of warfare at sea during the era of Nelson may be seen in the statistics of the battle of Trafalgar. The United Kingdom had 27 ship-of-the-line to oppose a total of 33 French and Spanish. The British suffered 1666 dead and wounded, including Nelson himself of course, and lost no ships. The combined French and Spanish losses were 5781 dead or wounded and an estimated 8000 captured. Twenty-one French and Spanish ships were captured, and while some were recaptured or sank in the storm that followed the battle, only one ship was sunk during the battle: a French vessel, the *Redoutable*. While ships could be disabled and captured and their crews slaughtered, they were difficult to sink. This would all change. In the Battle of Tsushima in 1905, the Russians lost 11 battleships and 8 cruisers sunk by gunfire or scuttled to avoid capture. The Japanese lost three torpedo boats.

Guns had evolved since Trafalgar. They now fired explosive shells, capable of penetrating the wooden hulls and setting them afire, making wood a most unsuitable material for warships. At first warships were "ironclads", wooden ships with iron plate bolted to their sides, but the iron ship was not long in proving its superiority, and remained the material of choice until cheap, mass-produced steel became available at the end of the 19th century. Steel not only allowed for stronger vessels but for larger ones as well. Iron was heavy and too flexible to allow ships to be built in excess of 300 feet in length.

Iron ships had added armor as well. The placement of the armor was dictated by the "vital" parts of the warship: its engines, magazine, guns and other areas necessary for its survival in battle. The gun mounts were varied as well. Some vessels still mounted guns in the traditional Nelsonian broadside configuration particularly in the earlier time of this discussion. The turret offered obvious advantages of field of fire and for large guns it was preferred, although as discussed in the main body of this work, the weight of turrets had drawbacks. Experiments with barbettes were common at first. This provided a fixed circular armored enclosure with a gun that rotated within it. The gun fired over the top of the barbette and might be provided with a shield attached to the gun for additional protection. Other hybridized solutions were tried such as the sponson, an enclosure that projected beyond the ship's hull with a slit opening through which the gun fired. Thus barbettes and sponsons were stationary armor with guns enclosed that rotated while the turret rotated with the gun.

The power behind the ships had shifted to steam before this discussion begins. While paddlewheels were used briefly, they were easily damaged in battle and proved to be less efficient than the screw propeller. Earlier in the development, ships were designated by their power train: PS for paddle steamer and SS for screw steamer. The propeller was originally shaped much like a screw, but was shortened to the shape we now see. The name persisted however. All the steam vessels presented here were powered by screw propellers. The steam engines that provided the power were initially of iron and therefore limited to lower pressure and efficiency than the steel engines that would replace them. A ship with a low-pressure iron boiler might be able to carry only 4 or 5 days' supply of coal to cruise under steam power at full speed. The early single stroke engine was replaced by the more efficient double reciprocal engine and the use of steel boilers allowed for greater pressure and therefore greater efficiency and range. Eventually piston engines would be replaced by turbines and then by diesel but not until after the period covered in this volume. Unlike the steam turbine which is inherently a horizontal machine, the piston or reciprocal engine is usually a vertical engine in the ships of this age. It often projected above the waterline making it more vulnerable to damage by enemy fire. This presented problems of armor protection, but more importantly, ships with iron engines had a relatively limited cruising range under steam power. All such ships were therefore equipped with masts and a full set of sails. The sails were usually stowed during battle to reduce the risk of fire, but the masts or poles served another purpose: communication. In the days before Marconi, no wireless transmission was available. Ships signaled to each other using signal flags flown from the masts, or occasionally by semaphore flags, lights or speaking trumpets as Nelson had done at the beginning of the century. It is worth noting here as well that ships could not begin to move without first "getting up steam", that is heating the boilers to appropriate operational temperature. The fuel on all ships at this time was of course coal.

A brief discussion of the power train and control of ships may be helpful. The steam engines were connected to the propellers or screws by a shaft, and in some cases several engines were linked together, which could sometimes be unlinked to conserve fuel. The engines determined the speed of the ship, the direction being controlled by the rudder or rudders that were controlled by the steering machinery located in the stern usually steam powered. The command center was the bridge where officers were located, but during battle, they communicated with the conning tower to "conn" or control the ship. The term "conn" probably comes from the word, control or possibly conduct. The conning tower was usually well armored while the bridge was provided with good visibility. The conning tower appeared on the first armored ships, *Glorie* and *Warrior,* and largely disappeared with the *Dreadnaught*. Even during this period the bridge was often preferred as the control center because of its greater visibility..

The ram as the primary weapon for sinking warships gained a prominence perhaps less deserved than should have been the case. The ram usually proved successful only in sinking vessels that were immobile. Evasive action and gunfire were sufficient to thwart the attack in most cases and there was often considerable damage done to the ramming vessel in such encounters as well. The only British battleship sunk by ramming, HMS *Victoria*, was accidentally rammed by her compatriot, HMS *Camperdown* during a fleet maneuver. The collision nearly sank the *Camperdown* as well. One reason the ram became so popular was that, as we shall see, armor was initially so superior to the guns that were available that a ship could ram an enemy vessel without suffering damage from the gunfire of her adversary. This situation persisted for only a brief time, much shorter in duration than did the prevalence of the ram. Many battleships were fitted with rams long after their advocates had abandoned them.

The construction of the guns was greatly influenced by metallurgy as well. Smooth bore muzzle-loading cannon had to be pulled into the ship in order to be loaded. The weight of the shot fired limited the size of these guns as well since the gun crew had to carry these projectiles to the muzzle and ram them to the breech for each load. The ninety seconds per volley that Nelson's crews could accomplish is all the more impressive when the difficulty in handling these monstrous weapons by hand is considered. If the shot or worse the gun itself were to break free of the restraining ropes, it became lethal: a "loose cannon".

As shells replaced solid iron shot, the guns themselves moved from smooth bore to rifled weapons. Rifled guns imparted a "spin" to the shell of approximately one turn for 40 calibers or 40 times the bore size of the gun. This spin increased range and accuracy. The British had many older smooth bores at this time, so the newer more serviceable ones were retrofitted by a technique developed by William Palliser. The cast iron barrel was bored out to a larger size and then a wrought iron rifled tube, stronger than the cast iron, was inserted. The nomenclature gave the size of the gun in pounds followed by RML (rifled muzzle loader as opposed to ML designating smooth bore muzzle loading guns) and cwt (British hundred weight or 112 pounds) designating the weight of the gun to distinguish it from other guns firing the same shell weight. Thus, a 64-pounder 71-cwt RML fired a 64-pound shell, weighed 71 cwt or 7952 pounds and was a rifled muzzleloader. The process used to convert the older ML, smooth bore muzzle loading guns, to the newer RML rifled guns, was termed the "Palliser Conversion". Palliser also developed a very effective armor-penetrating shell.

The obvious advantages of a breech-loading gun made its pursuit a paramount objective. The primary difficulty was that the breech of a weapon is subjected to the maximum stress, the full force of the explosion of the propelling charge. Iron was simply not strong enough to withstand this repeated force in early breech-loading designs. The first breech mechanisms were termed screw breechloaders and consisted of a hollow screw breech though which the shell and charge were inserted into the barrel of the gun. A block was then placed between the screw and breech and the screw was tightened against it. They were prone to disastrous failure and were soon abandoned. Eventually, when steel and more precise machine tolerances became possible, the sliding-block breech became available and was widely adopted. The sliding-block was an interrupted screw breech of varying design that could be swung into place and tightened by a quarter or third turn. The seal achieved was superior and not dependent on personnel strength, and of course was faster to use. William Armstrong, Thorsten Nordenfelt and several others contributed to this development.

This history is best dealt with in the text and context of the battles discussed, but a word about the gun size nomenclature first. Smaller guns were still designated by shell weight (3-pounder etc.) but larger guns were designated by the size of the bore and this was given in inches. The units are given in English measurement, inches and decimals perhaps in recognition of British dominance at sea and in her shipyards where half the world's warships were built. This explains the apparent odd sizes of some guns as seen in the text: i.e. 4.7 inch. A successful design might be placed on numerous ships. If the design was widely used a "Mark" or "Type" followed by a number was placed after the gun size. Also, when it is necessary to give the length of the barrel, this is done by placing a "caliber" after the bore size. The caliber is then multiplied by the bore size to give the barrel length. Thus a 16-inch, 50-caliber gun, designated 16"/50, is 16 inches in bore size and 50 times 16 inches in length: 800 inches or 66.6 feet, from breechface to muzzle. A longer barrel added weight and other restrictions to ship design, but increased a gun's range and accuracy. Incidentally, these were the guns that were placed on the Iowa class battleships in 1943, the last battleships built for the US Navy. The 16"/50 Mark 7's had a maximum range of 24 miles and at that range it took more than 90 seconds for the shell to reach its target. A ship traveling at 30 knots would cover 3/4 of a nautical mile in that time. Aiming such guns also had to make allowance for the motion and curvature of the Earth, the Coriolis Effect included. This could be obviously complicated during a ship-to-ship exchange, but against stationary land targets, which proved to be an increasingly common role for these vessels, the added range was a great advantage.

By the latter part of the nineteenth century, steel had replaced iron however, and the breech-loading guns dominated. The smaller caliber weapons, up to six inches in size were quick fire, termed "QF". The propellant charge and the shell came as single unit resembling a large "bullet" similar in appearance, but not in size, to that which would be loaded into a hand or shoulder weapon. This allowed a more rapid rate of fire and it was proposed that this would produce more damage than the slower rate of larger guns. Again this is discussed at length in the text and against the background of the battles fought. Various machine guns and Gatling guns were introduced during this period as well. At this time guns were aimed and fired by gunners stationed at or near the gun, not at a central fire control station.

An additional term should be discussed as regards the guns and ammunition used. The main supply of ammunition, both shells and propellant was stored below deck in one of the ship's armor protected magazines. During action a small amount of both shell and charge were at the gun itself so as to allow uninterrupted fire without undue risk. This was referred to as "ready ammunition" and was replenished as needed during action. It should also be kept in mind that the shell size and weight increases as the cube of the diameter. A 12-inch shell is not twice the size of a 6-inch shell, it is the cube of 2 or eight times larger.

Two final weapons deserve a brief discussion, the torpedo and the mine. These weapons evolved dramatically during this period and eventually morphed their names as well adding to the general confusion. At the outset of this book the torpedo was an explosive charge mounted on a pole that was driven into the bottom of a river or shallow harbor. The pole was pointed toward the expected path of enemy vessels and the charge was fitted with percussion detonators such that when a vessel ran onto it, the torpedo would explode sinking or damaging the enemy. It was this device that prompted Admiral

Farragut's famous remark during his attack in the harbor of Mobile, Alabama, the last major Confederate port on the Gulf of Mexico to be captured during the American Civil War. When, on August 5, 1864, he was warned that Mobile Harbor had torpedoes and that he should proceed slowly he replied: "Damn the torpedoes. Full speed ahead!" (although other less dramatic versions are quoted as well). By this expediency, he ran past the far more dangerous Forts Gaines and Morgan guarding Mobile with the loss of only one ship to a torpedo.

But torpedoes changed quickly from stationary obstacles to explosives delivered by small boats aptly termed torpedo boats. The explosive was still mounted on a pole and projected from the bow or sometimes towed behind the vessel. Delivery of the torpedo however relied on the enemy remaining inactive, which was seldom the case. Robert Whitehead took on the problem in the 1860's and developed a torpedo that could propel itself to an enemy ship, termed the "locomotive torpedo". These were initially quite slow, 6 to 10 knots and slower than the ships they were fired at. They were eventually refined and in future wars would be launched from airplanes and submarines to become responsible for more sunken tonnage than any other weapon. That was their future however, and in this work they were less effective as we shall see.

Meanwhile the stationary explosive became known as a mine, a name probably borrowed from the explosives placed in tunnels dug under fortifications on land. There were some early instances of explosives placed in harbors and electrically detonated from observation positions on shore, but eventually these devices became explosives charges with flotation chambers and were suspended below the surface from cables anchored to the ocean bottom, usually termed moorings. This allowed them to be deployed in deeper water as well. Percussion detonators were affixed on their surface and ships striking these were sunk or damaged by the subsequent explosion. So it was that mines more closely resembled the original "torpedo" with its fixed position, and the torpedo became a self-propelled device.

In closing, I would like to briefly mention the battle formations used. In the days of wooden ships, the line of battle was the line ahead. Each ship followed the ship ahead of it thus presenting a line with the broadsides of each ship facing to the sides of the line. With the advent of the turret and the ram there was a brief ascendance of the line abreast as an alternative. In this case the ships were next to each other, abreast or sometimes in an echelon or shallow "V" formation. This allowed for ramming, although it also limited the guns that could be fired at the enemy formation since none of the stern or side-mounted guns could be brought to bear. The line abreast also allowed a well-disciplined fleet to turn together and form a line ahead in either direction. When ramming was a popular weapon, the line abreast was a popular formation. With the ascendancy of the gun as the preferred weapon, this formation lost its popularity.

It is hoped that this brief discussion will add to the understanding of the ships and battles presented. Read on for the excitement and enjoyment of the drama from another age, but keep in mind the lessons for our own time and our own "battleships" whatever they are and whatever they become.

A brief aside regarding the British shipyards: It is this instrument of Britain that allowed her to dominate both the navies and the construction of the navies of the world.

Britain had a de facto policy during this time referred to as the two power standard: the United Kingdom would have a fleet equal to or greater than the next two largest fleets of foreign nations. This was made an official policy by passage of an act of the Parliament of the United Kingdom on May 31, 1889 termed The Naval Defense Act of 1889. Funds were provided to construct 10 new battleships and 34 cruisers as well as several other warships over the next 4 years. This provides ample evidence of the determination of Her Majesty's Navy to maintain naval supremacy, ("Rule Britannia! Britannia rule the waves!" wrote James Thomson in 1774). This act also emphasizes confidence in the capacity of British shipyards to produce the necessary ships to fulfill this policy. This they did.

Also of interest is the fact that in 1889, the two next largest navies belonged to France and Russia. In 1922, in the Washington Naval Treaty, Britain agreed to parity with the United States, while Japan was set at 2/3 of Britain. France and Italy were 1/3 of her naval strength and Russia was not a participant.

Rule, Britannia! (the refrain and first stanza)

By James Thomson

Rule, Britannia! Britannia, rule the waves!
Britons never, never, never shall be slaves.

When Britain first, at heaven's command,
Arose from out the azure main,
This was the charter of the land,
And Guardian Angels sang this strain:

Rule, Britannia! Britannia, rule the waves!
Britons never, never, never shall be slaves.

This is phrased as "heaven's" command. The angels commanded Britannia to rule the waves. The reader may wish to compare this to Kimigayo, the Japanese national anthem, shown on page 181 in the text.

THE BATTLE OF CHERGOURG, JUNE 19, 1864

INTRODUCTION

This encounter is really outside the scope of this text but I will mention it briefly here in order to emphasize an important issue.

The CSS *Alabama* was built in Birkenhead, England near Liverpool and served as a Confederate States commerce raider during the American Civil War, being very successful in that role. She was a wooden screw-driven, unarmored vessel mounting six 32-pound smoothbores in broadside, one 110-pounder Blakely rifle in a pivoted mount forward and one 68-pound smoothbore similarly mounted aft. She carried 170 men. She was built in Britain and sunk off the coast of France. She was possibly the most successful ship in the Confederate Navy even though she never docked at a Confederate port during her two-year career.

The USS *Kearsarge* was a screw-driven sloop-of-war launched in 1861 at the Portsmouth Naval Shipyard in Kittery, Maine. She mounted two 11-inch smoothbore Dahlgren guns, four 32-pound guns and one 30-pounder Parrott rifle. She carried 140 men. The only real difference between the two antagonists was that *Kearsarge* had chain cable armor in tiers along her midsections, but this was to prove a crucial difference.

An interesting piece of trivia regarding the Portsmouth Navy Shipyard is that it was named for Portsmouth, New Hampshire but was actually located across the Piscataqua River in Kittery, Maine. The actor, Steve McQueen, spent 41 days in the brig there in 1947 before he straightened himself out, embracing the US Marine Corps' ethic of discipline and self-improvement. He saved the lives of five other marines when their tank fell through the ice on an arctic exercise and was appointed an honor guard on President Harry S Truman's yacht. He died in 1980 of mesothelioma, a form of lung cancer associated with exposure to asbestos, which was commonly used in shipyards during the time he was at the Portsmouth facility. His acting accomplishments need no repetition here.

THE BATTLE

The *Alabama* entered the French port of Cherbourg on June 11, 1864, with the intent of effecting repairs and refit. The *Kearsarge* arrived a few days later and Captain Semmes on *Alabama*, fearing that reinforcements might soon arrive, decided to exit the harbor and engage *Kearsarge*. The French escorted *Alabama* to a point outside the three-mile limits of French sovereignty.

The battle then ensued with *Alabama* firing first while Captain Winslow on *Kearsarge* held his fire until the range was a more acceptable 1000 yards. The two ships circled each other on opposite tacks exchanging shot and while *Kearsarge* was hit, her "armor", minimal though it was, protected her from significant damage. *Alabama* received several hits, many at or below her waterline and after an hour's exchange she began to sink. She lowered her flag and finally brought the action to an end when one of her crew waved a white flag by hand (a table cloth it is sometimes reported). Captain Semmes requested aid in evacuating *Alabama* which was provided by *Kearsarge* along with a British yacht, *Deerhound*, and several French pilot boats. The *Alabama* suffered about 40 dead and 70 captured by *Kearsarge*. The remainder, including Captain Semmes, were taken to Southampton, England by the *Deerhound*, to the great anger of the crew of *Kearsarge* who felt they were rightfully their prisoners. Captain Winslow apparently personally prevented his crew from firing on *Deerhound*. *Kearsarge* suffered only one dead and two wounded.

Captain Semmes stated later that he would never have engaged *Kearsarge* if he had known of her armor and this is the point worth emphasizing. Wooden ships did not do well against the ordinance available at this time, while even the most minimal "armor" could withstand its fire. It is also true that the *Alabama* fired more rounds but with less accuracy, while *Kearsarge*'s fire was slower, more deliberate and better aimed, but this does not explain the huge disparity in casualties and damage suffered by the two vessels. The battle tactics were similarly unimaginative and both ships were hit several times, but *Kearsarge* suffered little while *Alabama* was seriously damaged, suffered significant casualties and eventually sank. The shell-firing gun had made iron armor necessary and that armor had rendered the guns ineffective -- at least temporarily so. What followed in the years to come was a duel between the gun and the armor each gaining ascendancy for a time and then losing it. That is the story told in this book and in truth is the story unfolding today, a "battle" between the offensive weapons and tactics and the defense arrayed against them..

TO THE BATTLE OF LISSA JULY 20, 1866

INTRODUCTION

The ironclads that engaged at Lissa were very similar to one another being basically armored versions of the three-masted wooden screw-driven frigates of the 1850s. Only *Affondatore* was distinctive, with two pole masts, tall twin funnels amidships, and a prominent ram bow; her two turrets were not at all conspicuous. While all the three-masted vessels were provided with a full spread of sail, we may readily understand that they fought under bare poles. Communication was by signal hoist or speaking trumpet. Guns were short-barreled muzzle-loaders; armor was of iron, with an average thickness of 4.75 inches. Optimum tactical range was considered to be 50-75 yards. Boarding parties were routinely included in the ships' action stations, and much was expected of the tactic of ramming. The lithograph purporting to illustrate this engagement that is represented in The Oxford Companion to Ships and the Sea is entirely erroneous and lacking in authenticity.

THE BATTLE OF LISSA

A certain amount of unavailable or contradictory data still plagues me with regard to this engagement, particularly concerning the numbers and identity of the wooden vessels present. The various accounts differ from one another, and occasionally differ in detail from the strength returns of the respective navies. With the exception of the Austrian ship, *Kaiser*, a screw-driven first-rate line-of-battle ship mounting 92 guns, the wooden ships took no significant part in the action, so this last information is of less significance than it might seem.

Austrian disasters on land and the rapid victory of Prussia made necessary a success at sea by Prussia's ally, Italy in order to sit at the conference table, (as Mussolini was later to say in another connection). The Italian government had been prodding the dilatory Admiral Persano to undertake some offensive operation. His procrastination having at length provoked a peremptory order direct from the king, he set out to capture the fortified Adriatic island of Lissa (previously under the rule of Venice before it was ceded to Austria in 1814. It is now Vis under the sovereignty of Croatia).

Arriving off the island in the afternoon of July 17, the Italian fleet pitted its 224 guns against the 88 of the Austrian works for the next two days. The bombardment is often called indecisive, although in fact it was a striking victory for the Austrian gunners. The ironclad *Formidabile* (sister to *Terribile*) was put out of action and obliged to make for home to effect repairs, thus being absent upon the day of battle. The Italian squadron as a whole suffered 16 men killed and 114 wounded in the course of the artillery exchange; also, the ships shot away much of their ammunition, *Re d'Italia* alone expending 1500 rounds. A landing attempt during Jul 19 was also driven off.

The third day, July 20 1866, preparations were being made aboard the ships of the squadron to resume the bombardment in support of a projected second amphibious assault upon the island. At 0750 the sidewheel dispatch boat *Esploratore* came down to report the Austrian fleet approaching from the north. Admiral Tegetthoff's squadron approached through the mist already arrayed in its line abreast battle formation, three consecutive double-echelon wedges of

18

seven ships apiece. The ironclads composed the leading wedge; *Erzherzog Ferdinand Max* held the central, leading position, flying Tegetthoff's flag. *Kaiser Max* was the central ship of the three in the starboard wing, and *Kaiser* led the second echelon of large wooden ships. I cannot determine the positions of the other vessels within the formation. The Austrians mounted a total of 74 rifles against 200 Italian (the designed armament of *Ferdinand Max* and *Habsburg*: 8.2-inch Krupp rifles, had not been delivered and they went into action mounting smoothbore 48-pounder shell guns).

It was an exceptionally misty morning, so it was not until 0920 that the Austrians were sighted standing down through the mist. Persano thereupon formed his ships up in line abreast; at 0950 he ordered simultaneous turn to starboard and line ahead, crossing the bows of the Austrian formation. Potter and Nimitz suggest that it was his intention to envelop the Austrian port wing. The Italian line was in the order (van to rear) *Principe di Carignano, Castelfidardo, Ancona; Re d'Italia, Palestro, San Martino; Re di Portogallo, Regina Maria Pia, Varese, Terribile* (Persano's flag in *Re d'Italia*).The Italian wooden ships, a considerable squadron, were ordered to remain 3000 yards to the south. In fact they never came even this close to the action, and participated not at all.

Then, as the newly purchased British-built turret ram *Affondatore* came up to starboard, Persano made what one of his countrymen has called "the most unfortunate decision an admiral has ever conceived". In the face of the enemy he halted *Re d'Italia* and transferred his flag to *Affondatore*, believing, as he said, that as commander he might most effectively oversee the action and exercise his judgment from outside the battle line. A squall blew up, and the change of flag was unobserved among the Italian squadron. Of greater immediate consequence, the halting of *Re d'Italia* threw the center and rear into some disorder and caused a considerable gap to open between the van and the center. Seeing this, Tegetthoff made "Ironclads ram and sink the enemy", and *Ferdinand Max* led the entire Austrian formation into the gap. The seven ironclads passed through the enemy formation, and the ships of the two wings peeled off in opposite directions to attempt to hold themselves between the enemy and the besieged island fortress. The Italian van had been commencing a turn

19

to port to envelop the Austrians, but seeing enemy ships to the rear Admiral Vacca in *Principe di Carignano* led the division on around and at 1045 opened fire. The cannonade quickly became general, and the mist, the billowing yellow clouds of gun smoke, and the coal smoke of steamships maneuvering at high speed combined to reduce visibility to a minimum. Consequently, the battle rapidly lost coherence.

Kaiser attempted to lead the woodens in the wake of the ironclads, but was obliged to wear away to starboard to avoid the Italian center, for *Re d'Italia* made for her but was unhandy enough that she could not close. *Kaiser* bore further to starboard, trying to make for the Italian wooden ships, but having eluded the presumptive, she was now confronted by the actual Italian flagship. So far from preserving his vantage point outside the line, Persano, foiled by the obscurity of the field, penetrated the melee, and *Kaiser* was now assailed by *Affondatore*.

Immediately after her brush with *Kaiser*, *Re d'Italia* had taken a shell right aft that disabled her rudder. Very imperfectly controllable, she was shortly obstructed by Austrian vessels, which, under the circumstances, her captain (Emilio Fas di Bruno) was unwilling to risk ramming. Attempting to back down, *Re d'Italia* was standing dead in the water when she was sighted by *Ferdinand Max*. The Austrian flagship had passed entirely through the enemy formation and was now doubling back into the fight. *Ferdinand Max* bore in at her maximum 11.5 knots, and *Palestro*, having thus far kept her station astern of *Re d'Italia*, comprehended the Austrian's purpose and attempted to divert her. She was thwarted by gunfire: one shell (from *Drache*, says Southworth) struck the wardroom and set her afire and the blaze proving impossible to contain, *Palestro* at length exploded and sank about three hours later at 1430. Her commanding officer, Alfredo Cappellini perishing in the blast together with all but 19 of her crew of 250.

The Austrian, *Ferdinand Max,* refrained from stopping engines short of impact, as was doctrine (to protect the engines from damage), and she struck *Re d'Italia* square amidships at full speed with a great tearing roar, smashing a massive hole into her hull below the waterline. *Re d'Italia* heeled far over to starboard under the impact.   As

Ferdinand Max backed clear, the Italian came upright and continued on over to port, capsized and sank. The doomed crew raised the cheer "Venezia è nostra!", ("Venice is ours!". The annexation of the Lombardo-Venetian Kingdom from Austria was one of the goals of the war for Italy, one that was ultimately realized at the conference table.) Fas di Bruno ended his life with his sidearm.

Persano, whom his command thought lost in the sinking, was in fact attacking *Kaiser* in *Affondatore*. (A minority opinion holds that it was, not the ship-of-the-line but the ironclad *Kaiser Max* which Persano now engaged. Potter and Nimitz refer specifically to "the wooden *Kaiser*" in this regard, and I accept their authority. The motive behind the alternate theory seems to be Italian patriotism rather than any desire to rehabilitate the lackluster and unfortunate Persano, who has very few apologists.)

At any rate *Affondatore* failed to ram, and was so damaged by fire from *Kaiser* that she was forced to draw off. There was scant respite for the Austrian however; her own formation had loosened badly, and she now confronted the Italian rear division looming up out of the smoke. *Re di Portogallo*, leading the four Italians, at once steamed in to attack the three-decker.

Captain von Petz, having insufficient sea room to evade entirely, sought to turn the tables and in fact succeeded in ramming *Re di Portogallo*. The glancing blow failed to damage the ironclad, however, while *Kaiser* lost her bowsprit and figurehead and was badly sprung forward. Further, the forward rigging either tore away with the sprit or fell slack, and the foremast toppled back against the funnel and caught fire. *Kaiser* lurched away and, passing astern of *Re di Portogallo*, at last penetrated the Italian line, albeit too much damaged to attack the Italian woodens, which still lurked well away to southward, out of the battle.

As the crew of *Kaiser* fought the flames, she was once more attacked by *Affondatore* as the turret ram passed back down through the battle zone. Despite her distress, the Austrian ship-of-the-line managed to evade the assault. It is maintained that *Affondatore*'s previous damage made her partially unmanageable at this time; and indeed she was so badly shot about that she sank in a storm off Ancona harbor on August, 6. (She was later refloated and served on until October, 1907.)

For the rest, the other ships cannonaded one another at close range, establishing that for the moment armor was entirely resistant to the best ordnance available. *San Martino* was briefly afire aft as the result of a shell hit. Although later alleging her own unfitness for further action, she seems not to have been much damaged. Various units tried, more or less clumsily and entirely without success, to ram the enemy. As described above, the major units of the Austrian fleet had now passed through the Italian line. *Kaiser*, still badly afire, shaped course for Lissa's harbor, and her division of wooden ships straggled after her. The Austrian ironclads fell back a little on the island in order to reform, and positioned themselves between the Italians and the beleaguered fortress.

Persano meanwhile had dashed down in *Affondatore* on the quixotic errand of bringing the Italian wooden ships up into the action while his fleet believed him dead. The Italians, disorganized, made no immediate effort to dislodge the Austrians from their advantageous position. As the fleets drew apart, firing gradually ceased; it had lasted less than an hour.

Admiral Vacca in *Principe di Carignano*, believing himself senior officer present after Persano's presumed death in *Re d'Italia*, gradually reformed the squadron (less *Palestro*, by now seriously afire, and *San Martino*, which steamed away southward making "Unfit for further action") and was on the point of leading a counterattack when *Affondatore* came up, Persano's flag at the main top. Persano, making the rather ludicrous "Give chase with freedom of maneuver," threw the fleet into a renewed confusion, which was not satisfactorily sorted out until nearly 1300.

At length the Italians once more stood in toward Lissa and the Austrian squadron, visual contact being made around 1320. Tegetthoff maneuvered in such a way as to keep his ships between the island and the enemy. The Italians commenced a desultory cannonade at rather long range, to which the Austrians did not reply. By 1430 Persano drew off once more, alleging in justification his incontestable ammunition shortage, and shaped course for Ancona. It is also true, however, that *Palestro* had just exploded with great violence and sunk without a trace in full view of both fleets. The effect of this on the morale of the Italian Fleet, and indeed on the morale of Persano personally, ought not to be underestimated.

Italian casualties were 620 killed (the majority in *Re d'Italia* and almost all the rest in *Palestro*) and 161 wounded. Austrian losses amounted to 33 killed and 153 wounded. Persano was court-martialed and though acquitted of cowardice and desertion, was judged guilty of negligence and incompetence, and was cashiered with the loss of all rank and benefits.

This battle, destined to be the last fleet action on the open sea for 28 years, is anomalous in the lessons it seemed to impart. The line abreast had triumphed over the line ahead; yet most subsequent theory and all subsequent practice reversed this judgment. The gun was transiently eclipsed and the ram acquired a largely spurious credibility that endured for many years.

THE BATTLE OF HAVANA, NOVEMBER 8, 1870

(It is only for the sake of comprehensiveness that I discuss this rather meaningless engagement at all.)

During the Franco-Prussian War, the gunboat *Meteor* was stationed at Havana in order to maintain Prussian interests in the Caribbean. She was a wooden vessel of 415 tons, completed in 1862. Her speed was about nine knots, and she carried one 24-pounder and two 12-pounder guns.

Early in November the French sloop *Bouvet* appeared in the area. She was a wooden ship dating from 1867, displacing around 760 tons, with a rated speed of 10.7 knots, and armed with one 6.4-inch and four 4.7-inch guns, The French commander issued a formal challenge to his Prussian counterpart, asserting that as their two nations were at war, it was fitting that they should meet in battle upon the open sea; that indeed it would be inconsonant with the honor of gentlemen to do otherwise.

The Prussian commander expressed himself in agreement with these sentiments, and it was arranged that the two ships should meet in battle on November 8. The people of Havana turned out in force to witness the engagement, thirty thousand by some estimates. The combatants steamed out to a point beyond the limit of Cuban territorial waters, and in full view of the city firing commenced. The tactics were of little interest; the two ships circled one another on opposite tacks, broadside to broadside and stern to bow, fighting just as *Kearsarge* and *Alabama* had fought off Cherbourg on June 19, 1864. In this case, however, neither was armored.

The gunnery was poor and little enough damage was done until at length a shot from *Meteor* ruptured the boiler of *Bouvet*. Wreathed in a cloud of white steam, the Frenchman hoisted sail and escaped within the three-mile limit. We may presume that the honor of gentlemen had been satisfied, at the expense of a few scalded engineers.

Meteor did not elect to pursue, but instead returned to Havana, taking up her berth once more to the acclaim of the populace. (The legal position here is questionable. While the Prussian's punctilious regard for international law in the form of the three-mile territorial limit must be commended, the equally well-established right of hot pursuit would have allowed a more resolute captain to compass the destruction of his foe without incurring the odium of his contemporaries.)

This proved to be the "major" naval battle of this war, the land campaign being the deciding area of conflict. Neither Prussia nor France had their navies in a state of preparation for the conflict, although France had a navy substantially larger than the meager force available to the German Confederation. There was some interference with commerce and the Prussian forces in the Baltic and North Seas are credited with preventing an amphibious invasion by France and thwarting an attempt by France to coerce Denmark into declaring war against Prussia.

NAVAL ENGAGEMENTS OFF THE PACIFIC COAST OF SOUTH AMERICA, 1864-1891

INTRODUCTION

(It may be helpful to consult the map contained in Appendix C, p 265)

These "banana wars" are now largely forgotten: Potter and Nimitz do not find them worthy of mention, and even the normally comprehensive Pemsel omits them entirely. In a sense, it is understandable that these incoherent battles fought in obsolete and worn-out second-hand warships should hold little interest for the military historian.

On the other hand, the technical student finds a good deal of interesting material. We see the true ironclad with its muzzle-loading rifled guns in action far more extensively than in any other part of the world; we catch an occasional glimpse of the locomotive torpedo in its infancy.

Also, and I daresay precisely because of the technical deficiencies of their equipment, we find that time and again the courage and initiative of individual officers and men substantially influenced the outcome of the struggle. We meet veritable heroes, men seemingly cast in the mold of an earlier age, and they are rather welcome. Therefore, in the final analysis we need not apologize for pursuing this neglected historical byway: it is a stirring tale of gallant sea fights, and eminently worthy of preservation on this account alone.

In view of the length of time under discussion, and of the fact that ships tended to remain in service through two or three wars, I have chosen to dispense with the customary introduction, and to speak of individual ships as we encounter them. Also, in the field of third-rate navies like these, we find Conway rather inadequate; on more than one occasion he maintains a position which is specifically contradicted by the entire remaining body of accessible literature. In such cases I have, not without hesitation, elected to "go with the majority." These battles span nearly three decades, but I have chosen to include the complete account together here for the sake of both continuity and simplicity.

HOSTILITIES WITH SPAIN, 1864-1866

Taking advantage of the United States preoccupation with their Civil War, Spain, who had never recognized Peru's independence had intensified diplomatic pressure upon her, much as France did in Mexico at this same time with Maximillian. A Spanish squadron was sent into the Pacific, comprising two wooden screw frigates, *Triunfo*, flying the flag of Admiral Pinzon, and the 38-gun *Resolucion*; also the 600-ton wooden screw gunboat *Covadonga* mounting two 70-pounder guns. On April 14, 1864, this force seized the Chinoha Islands, twelve miles off Pisco province. The government of Peruvian President Juan Pezet actually concluded a treaty with Spain subsequent to this occupation, but the strength of opposition to the settlement both within Peru and in neighboring countries was so great that President Pezet was deposed. His office devolved upon the more militant President General Prado, whose government at once repudiated the pact.

It was at about this time that Pinzon's flagship, *Triunfo*, was destroyed in an accidental fire. For reasons that are not clear, Pinzon was recalled to Spain in the aftermath of the conflagration, and Admiral Pareja was ordered to replace him. In view of the deteriorating situation in Peru, Pareja was dispatched to his new command with his flag in the 4878-ton wooden screw *Villa de Madrid*; the brand-new sister frigates *Regina Blanca* and *Berenguela* sailed in company.

In the meantime, at the instigation of the government of Chili, a four-power pact including Peru, Bolivia and Ecuador had been formalized for the purpose of resisting Spanish aggression. The arrival of Pareja with his reinforcements induced the allies to act, and a joint declaration of war against Spain was uttered on January 14, 1866. At this time the navy of Chili comprised only *Esmeralda*, a wooden screw corvette completed in 1854. She displaced 850 tons and was armed with twelve 40-pounder guns. Peru had just purchased two wooden screw corvettes, *Union* and *America*; these ships had been built in France (as *Georgia* and *Texas* respectively) to the order of the Confederate States Navy, but their hard-pressed purchasers had never contrived to take delivery. With the fall of Richmond the two had been placed on the open market. They were ships of 1150 tons and armed with 70-pounder and 9-pounder guns.

In response to the declaration of war, Admiral Pareja established a blockade of Valparaiso, correctly judging this to be the main commercial port of the allies. Within a few days, *Esmeralda*'s captain, Juan Williams Rebolledo, took his ship out of Valparaiso intending to take advantage of the isolated blockade station of the small *Covadonga*, the only vessel of the enemy that he could hope to master. He succeeded in surprising her, and in a brisk action, his crew boarded the Spaniard and took her. *Esmeralda* and her prize put into Coquimbo before the Spanish frigates could intervene.

This blot upon his honor so affected Admiral Pareja that he ended his own life; his body was cast into the sea without ceremony. Now thoroughly exasperated, the Madrid government dispatched Captain Casto Mendez Nunez to assume command, his flag in the two-year-old broadside ironclad *Numancia*. Her thickest armor was 5.5 inches, and she carried 500 men. "Spain was one of the few naval powers of the second rank to build broadside ironclads" as Conway remarks and doubtless it was reasoned, as so many politicians have vainly reasoned since, that the mere presence of so formidable an engine of war would sufficiently cow the foe that a favorable arrangement might quickly be agreed upon. (*Numancia* was incidentally the first ironclad to circumnavigate the globe and was under the command of Captain Mendez during that voyage, 1865 – 1868, leaving Spain for these encounters and proceeding to complete her circumnavigation to Spain.)

Immediately upon his arrival, Captain Mendez organized an attack upon the allied "combined fleet," which was at Abtao and comprised *Union*, *America*, *Esmeralda* and (gallingly) *Covadonga* with a Chilean crew. The action took place on February 6 1866. Fighting continued for two hours, the Spanish attack being "beaten off" by the allied warships. It is to be regretted that no creditable account of the details of this action is available. We may, as in most cases of this sort, console ourselves with the assumption that, since no ships were destroyed and no records of damage or casualties exist, the fighting was most likely simply a long-range artillery exchange of little tactical interest.

Having failed in his assault upon the warships of his enemies, Captain Mendez now resolved more vigorously to disrupt their commerce. On the morning of March 31, 1866, the blockade of Valparaiso gave way, to a bombardment of the unfortified city. Firing continued for three hours, the Spanish gunners concentrating upon the customs house sheds, which were known to contain quantities of valuable goods. The sheds, as well as other areas of the city, were consumed by flames; the value of the destroyed merchandise was set at 14,000,000 gold.

The Spaniards now steamed away and ceased to molest the Chilean coast. On the morning of April 25, they appeared off Callao, the port of Lima. The following day Captain Mendez issued a formal notification to all foreign nationals through their consulates, advising them that neutrals had four days to remove themselves and their property from the vicinity, for at the expiration of that time it was his intention to attack the fortifications that had protected the harbor since colonial times.

President General Prado arrived in the interim to take personal command of the defense; the members of his cabinet accompanied him, we may imagine somewhat reluctantly. About 1200 on May 2, the Spanish squadron entered the harbor and stood in toward the works. *Numancia* fired the first round, the guns of the Merced Tower replied at once, and the engagement quickly became general. By 1300 *Villa de Madrid* had been disabled, and was towed from the harbor. At about this time *Berenguela*, without apparent justification, alleged herself "in sinking condition." On the other hand, a shell from *Numancia* penetrated the Merced Tower and burst between two guns, detonating the ready ammunition. It is gratifying to be able to record that the minister of war was among those slain.

At 1700 the Spanish vessels ceased firing and withdrew. The carnage had been considerable: Spanish casualties were 40 men killed and 200 wounded, Peruvian, 200 killed and 500 wounded. I think the round numbers justify us in assuming the reported casualty figures to be estimates. The Spaniards accounted the bombardment a success, as they had substantially damaged the Peruvian fortifications, which was their sole intention, no landing or other offensive operation having been contemplated. The Peruvians held that they had successfully repulsed a Spanish attack and inflicted significant damage upon the attacking force.

No further operations were undertaken, and on May 9, Captain Mendez issued a decree stating that, as "grievous and sufficient punishment" had been inflicted upon the allied states for their affronts to the Spanish government, the honor of Spain was now satisfied, and the blockade of Valparaiso was forthwith "raised and nullified." It may be added that his ammunition was running quite short, and he lacked the manpower to organize any sort of significant landing force.

On May 12, 1866, the Spanish squadron sailed for home, and hostilities perforce came to an end, as the allied nations lacked the wherewithal to mount an attack against the Spanish homeland. A technical state of war endured until 1871, when with US mediation, a treaty was at length concluded, recognizing Peruvian sovereignty and terminating the war between Spain and the allies.

Still another unstated reason for Captain Mendez's rather urgent desire to be gone was the anticipated arrival of two ironclads that had been ordered by Peru in England, and which were presently in transit. Mendez greatly doubted the ability of his squadron to cope with the two ships, and we cannot much blame his diffidence.

Independencia was a conventional broadside ironclad; her thickest armor was 4.5 inches, and she carried 250 men. *Huáscar*, on the other hand, was a warship of the most advanced design. She carried her turret on the main deck, but to ensure adequate freeboard in a seaway she was equipped with hinged armored bulwarks amidships. These were lowered in action, giving her turret an excellent field of fire and incidentally doubling the thickness of her belt armor at the crucial time. Her belt and bulwarks were of 4.5 inch thickness, her turret of 8 inch. Her crew was of 170 men. Both she and *Independencia* featured a reinforced ram bow.

At about the same time Peru acquired from the United States two Canonicus class monitors which had been completed for the U S Navy, but which due to the conclusion of the Civil War were never formally commissioned into it. Originally *Catawba* and *Oneota*, they entered Peruvian service as *Atahualpa* and *Manco Capac* respectively. Their side armor was 5 inches thick, their turret armor, 10 inches; they carried 100 men.

In 1875 the Peruvian navy acquired *Pilcomayo* as well. She was British-built, one of the first ships to mount the new Armstrong breech-loaders. The 5.8-inch were 70-pounders, the 4 inch, 20-pounders. She was unarmored and carried 150 men.

THE BATTLE OF ILO, MAY 29, 1877

The perennial Peruvian revolutionary Nicolas Pierola having fomented his third unsuccessful insurrection, the government (we may think with considerable forbearance) exiled him to Chili. In May 1877 certain adherents of his, led by two brothers, naval lieutenants by the name of Carrasco, seized control of *Huáscar* and put to sea under cover of darkness. Pierola, forewarned of this escapade, had made his way to a Bolivian port, where *Huáscar* in due course took him aboard. It was the plan of the revolutionaries to return to Peru, put into whichever port seemed most favorably disposed, rally Pierola's adherents and seize control of the state. The news of Pierola's return, however, was received in Peru with indifference, and the anticipated supporters were nowhere to be found.

Huáscar's crew thus found themselves "men without a country," and food and fuel were running short; on the other hand, they were possessed of the most powerful warship in the Western Hemisphere. An attempt to requisition needed supplies at Pisagua was rebuffed, and in retaliation, *Huáscar* bombarded the town. There was nothing like enough manpower to land a force and seize provisions, so apart from revenge the cannonade served no purpose. It did, however, provoke the wrath of the government, and *Independencia*, supported by *Union* and *Pilcomayo*, was sent to seize or destroy the rebel. They ran her to ground and fought her for an hour and a half, but failed to either seize her or destroy her, or to even prevent her escape. Again details of the action are totally lacking, so I suspect another excessively cautious long-range duel of no particular interest. (It will be noted that the corvette *America* was absent on this occasion. Indeed, we will encounter her no more: Conway lists her as "discarded by 1880," and it may be presumed from the failure to utilize her at this time that if she had not yet been scrapped, then at the least her unseaworthiness must already have been manifest.)

The government in Lima thereupon determined to seek the aid of the Royal Navy, units of which, commanded by Rear Admiral de Horsey, his flag in *Shah* and with *Amethyst* in company, reposed presently at Callao.

Shah was one of a class of three very large unarmored iron frigates designed in 1866 although she herself did not complete until ten years later. She mounted sixteen 7-inch muzzle-loading guns and eight 5-inch breech-loading guns in broadside with two 9-inch muzzle-loading rifled guns mounted as bow and stern chasers along with various machine guns, Gatling guns and torpedo launchers.. She was an excellent sailer, logging 13.5 knots under canvas. She carried a crew of 600 men. *Amethyst* represents a type of warship that the Royal Navy commissioned in sizable numbers. They were designed for patrol work, service on distant stations, and "showing the flag" (or "gunboat diplomacy," the misnomer by which it later came to be known). Her particular class of five corvettes were launched in 1873-4 and completed in 1873-5. The guns were mounted in broadside: Twelve 64-pounder 71-cwt RML guns and two 64-pounder 64-cwt RML. (The nomenclature is discussed in the General Introduction section at the beginning of this work page 8, the Palliser conversion). The crew was of 225 men.

Then as ever, the Royal Navy was extremely loathe to become involved in shooting wars which did not directly affect British interests. Thus the facts of the case were quite carefully presented to the Rear Admiral: *Huáscar* had been declared "piratical" by the government in Lima after her bombardment of Pisagua; she had since forcibly halted mail steamers and removed persons and stores from them (which at least with regard to provisions was incontrovertibly the case, for the rebels had adopted this sorry expedient to replenish); the Peruvian Navy regretted to admit that it was powerless to end these lawless depredations. In the circumstances, de Horsey judged that *Huáscar* was indeed "a grievous and immediate hazard to navigation and trade," the sort of thing with which he was empowered to deal at his own discretion.

Confirmed in his opinion by the British charge d'affaires at Lima, to whom he had taken the trouble to submit the matter, de Horsey set out with his two ships to intercept and subdue the "pirate". She was sighted on the afternoon of May 29, 1877, off the port of Ilo. De Horsey demanded her surrender, with a five-minute ultimatum. At the expiration of this time, the "pirate" having failed to respond, a blank charge was fired, followed by a shotted gun fired across

Huáscar's bow. As "Pierola's flag remained flying" (precisely what flag he flew we are not informed, but it was most likely that of the Republic of Peru, of which this astonishing fanatic never ceased to regard himself as the rightful head), action commenced at 1506.

Huáscar at once stood in toward Ilo, continuously receiving and replying to British fire. *Shah* and *Amethyst* prosecuted the action at ranges varying from 2500 to 1500 yards. De Horsey regarded a range of 1000 yards to be optimum for his armament, but *Shah*'s unhandiness constrained him from closing for he doubted her ability to evade an attempt by *Huáscar* to ram at such a distance. Further, as *Huáscar* maneuvered continually, marching and countermarching before Ilo, the British were frequently obliged to cease fire in order not to endanger the town with their misses.

After two hours of this frustrating impasse, *Amethyst* was ordered to stand in to the shore and attempt to force *Huáscar* out into deeper water. For whatever reason, the rebel promptly accepted the hint, and turning directly toward *Shah*, ran up upon her at full speed. *Shah* held the range open as much as she could, the "pirate" bearing after her with evident intent to ram. At this angle *Huáscar*'s forecastle obstructed her turret, but the two exposed 40-pounders were firing shell. *Shah*'s Gatling machine-guns were trained on the 40-pounders, driving their crews to cover and thus silencing them; 64-pounder shells stripped away *Huáscar*'s rigging and unarmored superstructure; and the 7-inch guns did what they could against her armored bulk. *Shah* succeeded in keeping clear of *Huáscar*'s ram, and rather than proceed to seaward of her, the rebel turned back toward Ilo. As she passed, *Shah* discharged a Whitehead 14 inch locomotive torpedo at her (the first ever fired in combat). *Huáscar* sighted the wake of the torpedo and turned directly away from it, and as *Huáscar* was going eleven knots, the torpedo, having only a speed of nine knots could not overtake the vessel. The torpedo's track was observed going direct about half the distance towards her. *Huáscar* continued to stand in toward the coast, where she was lost to sight in the gathering darkness, and firing ceased at 1745.

Shah and *Amethyst* stood watch outside Ilo, and at 2100 *Shah* launched her steam pinnace with an outrigger (spar) torpedo, together with her whaleboat carrying another 14-inch Whitehead torpedo. The two made their way into the harbor but failed to find *Huáscar*, and at length returned without incident to their parent ship.

In the morning, it was verified that *Huáscar* had slipped away under cover of night. At length she was discovered at Iquique, Peru, where Pierola had surrendered her to the Peruvian authorities rather, as he said, than see her fall into the hands of the British. Received cordially at first by the Peruvians, the British were permitted to put *Shah*'s gunnery lieutenant and a lieutenant of the Royal Marine Artillery aboard *Huáscar* to assess the effect of their fire. (Despite *Huáscar*'s enthusiastic gunnery, neither of the British ships had been struck. Her guns were undermanned and she managed only 40 shots, obviousky not well aimed) The team reported that *Huáscar* had sustained a total of seventy to eighty shell hits. Her upper deck was a shambles, the funnel casing and bridge repeatedly perforated and the boats and masts virtually swept away. In addition, it was specifically noted that two 64-pounder shells had penetrated the armored bulwarks and dented the plate beneath (in each case the bulwark itself was gone, blown free of its hinges by the violence of the explosion); that a 7-inch common shell had penetrated three inches into the turret's eight-inch plate; and that twelve rounds from the Gatling guns had passed completely through the funnel. Certainly the two 64-pounder penetrations, and probably the 7-inch and the Gatlings as well, occurred when *Huáscar* in the course of her attempt to ram came within virtually point-blank range of *Shah*.

Confirmed casualties aboard *Huáscar* were one man dead and one wounded. There were rumors of many more casualties, however; indeed, rumor of all sorts was rife in Iquique, for while Pierola and his officers were held prisoner, by some bureaucratic oversight (or through the efforts of some fellow conspirator in high office) the men of the rebel crew were not detained by the authorities. Their highly-colored accounts of the previous day's action quickly "made the rounds," and a spontaneous outburst of anti-British sentiment swept the population. The mood turned ugly, and I daresay, Her Majesty's officers were glad enough to get back aboard without incident. The one dead man was buried beneath an extravagant monument bearing, among others, the inscription, "This gallant hero gave his life fighting for his country against the British." The memorial was purchased with funds that had been raised by public subscription in a matter of hours.

In the ensuing days an anti-British mania swept the country. There was brief talk of an official protest by the government, but the British charge d'affaires at Lima (who it will be recalled had been privy to the entire incident almost from the beginning) intimated that the British response might well point up facts which, in the present climate of public opinion, would be politically disastrous, and no protest ever materialized.

As for Nicolas Pierola, not only did he escape with his life, but since in the circumstances any prosecution of him would have been impolitic in the extreme, no legal proceedings whatever were lodged against him, and he was de facto permitted to go at large in Peru. In fact, by the inscrutable permutations of the mass mind he quickly came to be regarded as a sort of patriotic martyr: it was largely by virtue of this whole preposterous fiasco that, as we shall see in due course, he at length attained to a travesty of his lifelong ambition.

NAVAL INCIDENTS OF THE (GREAT) PACIFIC WAR 1879-1884

INTRODUCTION

This was a struggle for control of the guano- and nitrate-producing provinces of Tacna, Arica, and Tarapaca (Peru) and of Atacama (Bolivia). Atacama was Bolivia's only outlet on the coast. Chilean companies engaging in exploitation of nitrate were so heavily taxed by Peru and Bolivia that Chile went to war.

Thus Dupuy and Dupuy, and for the present purpose I cannot improve upon their general introduction. (I will confess that I hold this particular war in special affection, not only for its bizarre battles and astonishing improvisations, but also as having been fought for the possession of bat shit.)

Bolivia possessed no naval force.

We are already familiar with the navy of Peru. In 1879 it comprised *Huáscar*, *Independencia*, *Union* and *Pilcomayo*. The two monitors were deployed as floating coast defense batteries, *Atahualpa* at Callao and *Manco Capac* at Arica.

Chile's navy had expanded greatly since we last examined it in 1866, largely as a result of a naval arms race with Argentina that continued until 1902 when Great Britain arbitrated a binding naval arms limitation pact. Both nations subsequently sold vessels to other powers including Japan where they saw service against Russia.

Pride of the Chilean fleet were two British-built sisters, central battery ironclads; *Almirante Cochrane* and *Blanca Encalada*, four years old in 1879. (Conway asserts that the latter ship was originally named *Valparaiso* and "renamed *Blanco Encalada* about 1890", but all other sources refer to her as *Blanco Encalada* throughout her career, and I shall do likewise.) The 9-inch muzzle-loading rifled guns were mounted behind embrasured ports, three to each side. These ships were far more maneuverable than the Peruvian ironclads because they featured twin screws, rather than a single screw inside the rudder such as equipped the older vessels. Armor was of 9-inch maximum thickness; complement, 500 men.

Two sister wooden screw corvettes dating from 1866, *Chacabuco* and *O'Higgins* are the subject of contradictory data. Conway fails to mention *O'Higgins* and gives *Chacabuco*'s displacement as 1101 tons and her horsepower, as 1200, while Akers credits the pair with 1670 tons displacement and 800 horsepower apiece. The main armament was three 8.2-inch (150-pounder) Armstrong breech-loading rifles, with a secondary armament of (Conway) two 70-pounder and four 40-pounder BL or (Akers) four 40-pounders. Their precise appearances cannot be determined with accuracy.

Magallanes was a barque-rigged corvette with one funnel; she completed in 1844 and displaced 950 tons. Her armament was one 64 pounder, one 7 inch, and two 4 inch guns. She carried 143 men.

Abtao was a wooden corvette, "barque-rigged, with a clipper bow and one funnel set well aft." She was of 1600 tons, armed with one 5.8 inch and four 4.7 inch guns. She completed in 1866 and carried a crew of 130.

Also still in service were the venerable *Esmeralda* and *Covadonga*, the former twenty-five years old in 1879 and the latter probably of much the same vintage.

It should be noted as well that the Chilean Navy was manned by professional seamen of long service, whereas Peru's ships were maintained by cadre "skeleton crews" in peacetime and recruited up to strength upon the outbreak of war by conscription of dockside laborers, for the most part Peruvian Indians.

INITIAL MANEUVERS

On February 9, 1879, Chilean troops embarked in ten transports and, escorted by the eight warships of the navy, were carried to Antofagasta, Bolivia. This place was occupied on February 14, without resistance by 500 men under the command of Colonel Sotomayor, and the surrounding country was quickly consolidated.

The navy instituted a blockade of Iquique, the main seaport of Tarapaca province. This point formed their "center of operations" during the ensuing months, despite the fact that the entire shoreline was in enemy hands. The ships set forth from and returned to the blockade positions while carrying out bombardments of land targets and suppressing coastal trade. The Peruvian fleet having shown no inclination to challenge these activities, the now promoted to admiral, Juan Williams Rebolledo set out on May 16, 1879, to reconnoiter Callao and assess the possibility of forcing action upon the enemy. He brought with him in company *Almirante Cochrane*, *Blanco Encalada*, *Chacabuco*, *O'Higgins*, *Abtao* and *Magallanes*, leaving *Esmeralda* and *Covadonga* to "maintain the blockade." In fact these latter ships were his least battleworthy units, obsolete, worn-out, poorly armed and slow, and it is clear that Williams' purpose was to leave them well out of harm's way while he engaged the enemy with his fully operational ships. Alas for his good intentions, it was to fall out exactly the opposite.

By a remarkable coincidence, it was also on May 16 that President General Prado of Peru departed Callao: it was his intention to reinforce (with three transports loaded with troops) and, as was his custom, to take personal command of the Peruvian Army which was assembled at Tacna. Escorting him were Captain Miguel Grau (senior naval officer present) in *Huáscar* and Captain Juan Guillermo Moore in *Independencia*.

The Chilean force moved northwards well out to sea, to escape observation from ashore; the Peruvians hugged the coast. Thus the two squadrons passed one another unobserved, coming within forty miles of one another at the nearest point. President General Prado and his troops were in due course landed at Arica without incident. At this point Captain Grau was informed of the situation off Iquique. He at once determined to proceed thither, destroy the vastly inferior *Esmeralda* and *Covadonga*, and raise the blockade.

THE BATTLE OF IQUIQUE, MAY 21, 1879

Huáscar and *Independencia* departed Arica under cover of darkness on the night of May 20, and arrived off Iquique around daybreak on the 21st. They were in visual contact with *Covadonga* and *Esmeralda* by about 0700.

Captain Arturo Prat of *Esmeralda*, senior Chilean officer present, conferred with Captain Condell of *Covadonga*. He delivered what has been preserved as a rather histrionic and overlong address, noting the honor of the Chilean flag, and proposing that despite the manifest inferiority of force the two wooden ships should fight to the end:

> "Lads, the struggle will be against the odds, but cheer up and have courage. Never has our flag been hauled down in the face of the enemy and I hope, thus, will not be this the occasion to do so. From my part, as long as I live, this flag will fly in its place, and if I die, my officers shall know how to fulfill their duties. Long Live Chile!"

"Fine with me," replied Condell, and the two men repaired to their respective commands to prepare to receive the enemy.

At 0800 *Huáscar* opened fire upon *Covadonga*, but as *Independencia* passed in front of her and made directly for *Covadonga* meaning to ram, *Huáscar* transferred fire to *Esmeralda*. As the action from this point devolves to a pair of ship-to-ship duels, it is most convenient to treat the two battles in sequence, although they occurred simultaneously.

Esmeralda returned the fire of *Huáscar*, and began to build up to full speed in order to maneuver as she had been at anchor when the enemy arrived. The strain was too great for her ancient boilers, however, and two of them burst in rapid succession, reducing her speed to around three knots. Prat limped his ship into shallow water inshore, to prevent the deeper-draught *Huáscar* from ramming, and the gun battle continued at 1000 yards. Peruvian gunnery was extremely inaccurate; the Chileans scored repeatedly upon *Huáscar*, but their obsolete 40-pounders lacked the power to significantly damage her.

Peruvian artillery began to take *Esmeralda* under fire from the shore and after sustaining a number of hits, with three men dead, Prat coaxed his crippled ship to seaward. Firing between the two ships continued unabated as the range came down to 650 yards. At this distance, a l0-inch shell struck *Esmeralda* near the waterline, the only Peruvian hit of the battle at anything greater than point-blank range. Casualties were inflicted, and *Esmeralda* was briefly afire.

Huáscar now bore in to ram her virtually immobile foe. She struck *Esmeralda* abaft the mizzenmast, but owing to the Peruvian having stopped engines too soon, the blow was a glancing one. *Huáscar*'s two 10-inch guns were fired into the Chilean ship at the moment of impact, working "terrible slaughter of (the) crew." At this same instant, Captain Prat, sword in hand, shouted the command to board and leaped onto the deck of *Huáscar*. As the roar of battle drowned his voice, and as the two ships parted so rapidly, only one man, Sargent Aldea, succeeded in following his captain. Members of the crew of *Huáscar* called upon Prat to surrender, pointing out the hopelessness of his situation, and Captain Grau shouted his personal offer of quarter. The Peruvians were genuinely reluctant to take the life of so gallant a foe, but the berserk Prat charged on unheeding along the deck of the ironclad, with Aldea close behind. Signal Lieutenant Velarde stepped out from the conning tower to impede their progress, but Prat shot him down with his sidearm. At this the patience of the Peruvians was exhausted, and Prat and Aldea died in a volley of rifle fire. ("Prat's self-sacrifice was not aimless, for Grau declared afterwards that, if Prat had been followed by his crew, they might possibly have taken the *Huáscar*, so demoralized were the crew of Peruvian Indians.")

The range remained virtually point-blank, *Huáscar* receiving many shell hits and repeatedly striking *Esmeralda* with 40-pounder and machine-gun rounds. *Huáscar* was preserved by her armor, but the Chilean crew suffered many casualties. At 1150 *Huáscar* rammed *Esmeralda* a second time, striking her at a 45° angle on the starboard bow. At the moment of impact Lieutenant Ignacio Serrano and twelve men boarded *Huáscar*, but were instantly cut down by a fusillade of rifle and machine-gun fire. *Esmerada*'s engine room and magazines were now flooded, she was dead in the water, half her crew was slain;

but she still floated, her flag flew at the maintop, and her guns still fired.

Huáscar now rammed her immobile enemy a third time, striking her full amidships and firing the two 10-inch guns into her at the moment of impact. At the same moment the last of *Esmeralda*'s ready ammunition was fired off into the foe, and two minutes later, at about 1200, *Esmeralda* sank with colors flying. Fifty men survived of her crew of 200.

Meanwhile, *Covadonga* had fled southward along the coast, as close inshore as she was able. The deeper *Independencia* could not ram, and as her guns were mounted in broadside, rather than turret-mounted as *Huáscar*'s she was unable to bring them to bear upon the enemy sharp off the port bow, but was obliged to follow a parallel course 200 yards further offshore and about the same distance behind *Covadonga*. On the other hand, the two pivot-mounted 70-pounders of the Chilean subjected *Independencia* to a persistent, accurate and galling, although largely ineffective fire, to which she was unable to reply.

Frustrated by his inability to come to grips, and with the morale of his crew deteriorating under the thunderous detonations of 70-pounder shell against *Independencia*'s plate, Captain Moore resolved upon a risky and precarious expedient. A lookout was stationed upon the extreme bow; visually estimating conditions immediately ahead, he shouted course and speed changes back, and thus in effect conned the ironclad from his unorthodox position.

Proceeding in this way, *Independencia* maneuvered closer inshore and succeeded in coming to within 200 yards of *Covadonga*. Her unusual system of navigation was not lost on the laconic Captain Condell, however, and he determined to exploit it. As *Independencia* came up to the particularly treacherous shoals of Punta Gruesa, ten miles south of Iquique, a Chilean marksman shot the lookout dead. That hapless individual tumbled into the sea, and the blinded ironclad blundered hard aground within thirty seconds. The sharp rocks punched through her hull and held her, while the current swung her around until she came to rest with her bow pointing out to sea, with a considerable list and shipping a good deal of water.

Covadonga turned back at once, maneuvered into position under the stern of *Independencia,* and commenced a close-range fire to which the Peruvian was utterly unable to reply. The relatively weak angled joints of the stern plating began to give way, and a fire started in the stern which the crew, by now well-nigh witless with terror, could by no means by induced to extinguish. Moore was on the point of hauling down his flag when *Huáscar* was seen bearing down from the north, her battle with *Esmeralda* at length concluded, and Condell finally esteemed discretion the better part of valor and made off toward Antofagasta.

(A History of Chile by Luis Galdames offers what I must regard as a deliberately garbled version of this battle, maintaining that *Independencia* was "cannonaded almost to the point of surrender" by *Covadonga* before she ran aground. The absurdity of this is patent; the motivation apparently Chilean patriotism, which it must be said is rather excessive, since a true account does the Chilean navy ample credit. Although Arturo Prat is esteemed the immortal hero of this day, I would be so bold as to suggest that in fact the more moderate gallantry of Captain Condell was of far greater service to his country.)

Captain Grau herded the survivors of *Independencia* aboard his own ship, and after a personal inspection of the grounded ironclad judged her unsavable and ordered her burned. Under the circumstances it was an irreparable loss. *Huáscar* was herself the worse for wear: her mast was so damaged that it had to be unshipped, her bow leaked rather badly after the repeated ramming of *Esmeralda,* and her turret was out of line, turning only with difficulty.

Nevertheless, Grau set his course to southward, and presently engaged *Covadonga* at Antofagasta. The gunboat returned a vigorous fire, which was augmented by three heavy guns ashore. A shell from one of the latter penetrated *Huáscar*'s armor. Damage was slight, but Captain Grau was not slow to draw the proper conclusions. His ship (now virtually Peru's only naval asset) was vulnerable to the guns ashore, and he made haste to withdraw out of their range, thus perforce leaving *Covadonga* unmolested.

The underwater telegraph cable linking Antofagasta to Santiago was dredged up and cut, such damage as could be accomplished quickly was done ashore, and *Huáscar* set out toward Callao to refit and put ashore some of the excess personnel inherited from *Independencia* (living conditions must have been extremely cramped).

On June 5, 1879, *Huáscar* sighted smoke on the horizon and Grau turned to investigate. He made an abrupt about-face upon recognizing *Blanco Encalada* in company with *Magallanes*; a few shots were fired at extreme range, but no hits were secured. The two Chileans took up the chase, but after eighteen hours, Grau outdistanced them and visual contact was lost. On June 7 *Huáscar* put back into Callao.

There was carnival in the streets of Lima as reports of the "great victory" of Iquique circulated, but we cannot imagine that Miguel Grau was much pleased: not even when his promotion to the rank of admiral was announced. (Upon receiving, in the field, the news of his promotion to the rank of field-marshal, Erwin Rommel remarked, "I would much rather he had given me two more divisions." While Grau may have shared something of this sentiment, he at least, unlike the German, had the cold comfort of knowing that his president had no more ironclads, and that only the honorific award was in his power to bestow.)

HARRYING THE CHILEAN COAST
JULY – SEPTEMBER, 1879

Huáscar was laid up for such repairs as could readily be effected, while Grau consolidated the worthwhile elements of the crews of *Huáscar* and *Independencia*. He gave his nearly worthless members what amounted to active encouragement to desert, and signed on as many experienced seamen as he could induce. His thirty Englishmen seem to have been the pick of the bunch, although I dare say they would have appeared a ragtag enough lot in the home island.

Admiral Grau was received by President General Prado upon reporting that *Huáscar* was fit for sea. Prado commanded him to interfere by all means with Chilean trade and military operations, but expressly forbade him to accept action with either of the two enemy ironclads.

Huáscar sailed at the beginning of July. In the course of the ensuing months, in addition to engaging in a succession of naval actions which we shall consider in detail, she took two merchantmen, destroyed numerous launches and coastal traders, and at various times bombarded and destroyed property at the ports of Carrizal, Chanaral, Huasco, Tocopilla, Taltal and Caldera.

On July 10, 1879, *Huáscar* attacked *Magallanes* as she maintained the blockade of Iquique. Gunnery on both sides was execrable, and three attempts to ram by *Huáscar* met with no success. The Peruvian was lining up for her fourth attempt when *Almirante Cochrane* was reported approaching at speed, and Grau made his escape in conformity with his orders.

In mid-July *Huáscar* was joined by *Union*, and on the twenty third of that month the troop transport *Rimac* was intercepted and taken. She carried a regiment of cavalry and 500 horses. This coup probably did as much as anything to outrage public opinion in Chile, where the continued failure of the navy to hunt down this lone marauder was creating a seething uproar. On August 17, *Huáscar* entered the harbor or Antofagasta and attacked *Abtao* and *Magallanes*. Disdaining his inaccurate guns, Grau discharged a wire-guided Lay locomotive torpedo at *Abtao*. The concept of this weapon was far ahead of its time, and the technology of the day was in nowise

45

adequate to support it. After running true part of the way toward *Abtao*, the device suddenly made a 180° turn and sped back toward *Huáscar*, completely out of control. Catastrophe was averted by the presence of mind of Lieutenant Diaz-Conseco, who dove into the water, swam up to the torpedo and succeeded in diverting it manually.

Huáscar now opened fire, and the two Chilean wooden ships, which had maneuvered close to shore, replied at once, supported by those shore batteries that Admiral Grau must surely have remembered. While Southworth. informs us that *Abtao* was reduced to "a badly battered wreck" in the course of this battle, I find little enough evidence to support such extravagant terminology. What is beyond dispute is that *Huáscar* once more withdrew from Antofagasta leaving her two opponents afloat.

In September, Grau put in to Arica, meaning to give *Huáscar* the overhaul that by now she desperately needed, and probably to scrape her bottom as well. President General Prado forbade the ironclad to be laid up, however, asserting (quite correctly) that she was the only impediment to a major invasion of Peru and Bolivia. The allied armies were not in condition to face such an assault, and Grau was given categorical orders to return immediately to sea.

Meanwhile, as has been mentioned, all Chile was in turmoil at the inability of the navy to contain, much less destroy, the rampaging *Huáscar*. In response to public clamor, the ships of the navy were brought into port and thoroughly overhauled. It was not necessary to relieve Admiral Williams of his command, for he (doubtless sensing the direction of the wind) retired from the navy for reasons of health. Rear-Admiral Galvarino Riveros Cárdenas was appointed in his stead.

Huáscar and *Union* put to sea late in September, 1879 in compliance with the president's command, with the greatest misgivings on the part of Admiral Grau. The two ships were reported south of Arica, and on October 1, the Chilean Navy put to sea. Riveros detached Captain Latorre, his flag in *Almirante Cochrane* with the armed merchant cruiser, *Loa* and *O'Higgins* in company, to patrol off Mejillones. He himself in *Blanco Encalada*, accompanied by *Covadonga* and the armed merchant cruiser *Matias Cousino*, cruised further to the south.

46

THE BATTLE OF PUNTA ANGAMOS,
OCTOBER 8, 1879

Proceeding northward in the vicinity of Antofagasta, *Huáscar* and *Union* early in the morning of October 8, detected smoke to the north. The Peruvians closed, identified *Blanco Encalada* as she rounded Angamos, and at once turned away and ran to seaward. *Huáscar* gained on her pursuers (for *Covadonga* also took up the chase), and Grau's chance of eluding them seemed good, when *Almirante Cochrane*, *Loa* and *O'Higgins* were sighted on the horizon, steaming in from the northwest to cut him off.

Ordering *Union* to make the best of her way out of the battle, Grau prepared to receive the enemy. The corvette made good her escape, steering to seaward and passing north of *Almirante Cochrane*. *Loa* and *O'Higgins* at once took up the pursuit, but at length *Union* outdistanced them and visual contact was lost without fire having been opened.

Of course *Huáscar* had no such good fortune; her time was up at long last, and everyone present knew it. *Huáscar* began to close *Almirante Cochrane* and opened the engagement at 0920, firing her 10-inch guns at a range of 3000 yards. The first round fell short; the second struck *Cochrane* squarely on the bow, but failed to explode. The third round was short, but the fourth ricocheted off the water, struck and penetrated *Cochrane*, doing no serious damage. (Thus the "chronic inaccuracy" of the 10-inch Armstrong ML rifles of *Huáscar* is placed in perspective: plainly what were needed were simply trained English gunners.) So slowly did the big guns fire that after four rounds the range had come down to 2000 yards, and *Cochrane*, hitherto silent, turned to fire a broadside. Hits were registered immediately, *Cochrane*'s fourth 9-inch shell striking the turret, inflicting twelve casualties and knocking the turret off its rollers to jam it. With the greatest exertion the Peruvians shoved the turret back into place and firing resumed as the two ships continued to close.

Two 10-inch shells slammed almost simultaneously into *Almirante Cochrane*, penetrating her armor and destroying one of her 9-inch guns. The Chilean began to lose way and appeared to be only erratically under control; the cause was simply the alarm and

47

consternation with which her crew responded to the damaging hits, but Admiral Grau seems to have believed her far more seriously injured than she was. *Huáscar* bore in to ram, but *Cochrane* evaded, and at 0955 at virtually point-blank range a 9-inch shell penetrated *Huáscar*'s conning tower and exploded within. Grau and his signal lieutenant were killed instantly; only the uniformed right leg of Admiral Grau was recovered for burial.

Blanco Encalada had meanwhile been making her best speed toward the enemy, but as she closed the maneuvering of the two combatants resulted in her fire being masked by her sister. After a near collision with *Cochrane*, *Blanco Encalada* passed under her stern and opened fire at last upon *Huáscar* from a range of 600 yards. The elderly but ferocious *Covadonga* had also come up and added the fire of her 70-pounders to the holocaust of point-blank fire which now engulfed the Peruvian.

Huáscar fought back as well as she might. Captain Elias Aguirre took command, but was almost immediately decapitated by a Chilean shell. The next in succession, Captain Manuel Carbajal was already too badly wounded to take command and so passed the office on to the next junior officer, Lieutenant Rodriguez, who was however instantly struck dead while still in his presence. Command then passed to Lieutenant Enrique Palacios.

By this time *Huáscar* was in pitiable condition. It was 1030, and the old ironclad was completely disabled. Her steam steering gear had been shot out; a shell penetrated the turret and exploded inside, destroying one gun and rendering the turret inoperable; the superstructure was demolished and even the armored hull had been holed repeatedly. Lieutenant Palacios was contemplating the disagreeable but increasingly inevitable prospect of surrender when he was slain by shellfire, and it was left to his successor, Lieutenant Pedro Garezon, to try to save what remained of his crew and to scuttle the ship. As he sent officers below to open the valves, however, members of his crew lowered *Huáscar*'s flag at 1040, and the Chileans at once sent boarding parties across in a number of boats. Lieutenant Garezon attempted to organize his men to repel the enemy, or at least delay them until *Huáscar* was beyond saving; but they refused, well knowing that if they offered resistance subsequent to the striking of the flag they would all be massacred,.

The Chileans came aboard without resistance. Their officers dashed below at once and succeeded in reclosing the valves. (Pike asserts that "just as the Chileans boarded her, the *Huáscar* sank", but this is plainly false. Doubtless Pike accepted at face value some patriotic Peruvian myth; the lie is the more pathetic since *Huáscar*, following a period of Chilean service, was preserved in that country as a museum, and survives to the present day.)

Still the ship was in precarious condition. She was afire in eight places, although only the conning tower was seriously ablaze, and there were three feet of water in the hold. Sixty-four of the 193 men aboard were either dead or wounded. The crew were made prisoners, and *Huáscar* was towed into Mejillones Bay. There her steering was set right, and she proceeded under her own power with a Chilean crew to Valparaiso. There she was repaired and in due course incorporated into the Chilean Navy. A commemorative plate was set into her deck to mark the spot on which Captain Prat had fallen.

Command of the sea having thus passed irrevocably to Chile, a landing was mounted in force at Pisagua. The combined Peruvian-Bolivian armies were defeated in heavy fighting at San Francisco on November 16. On the twenty seventh the Chileans were repulsed before Tarapaca, but the allies were too weakened and disorganized to follow up their success. President General Prado recognized this brief respite for what it was; that canny and realistic old soldier well knew that the game was up, and made haste to cash in his chips. At the beginning of December, 1879, he handed over to his vice-president and departed, with as much of the treasury as he could secure, to "raise a loan and buy ironclads" in Europe.

The true nature of Prado's desertion was not lost on the population; the troops in Lima mutinied against his nondescript successor. and the people proclaimed revolution. At this moment, prosecuting the last of his "hare-brained seditions", Nicolas Pierola came once more to the forefront of events. There could be no question of his personal valor, and his patriotism was proverbial; he was appointed by universal acclamation Supreme Chief of the Peruvian Republic. The ambition of this remarkable man was thus at length fulfilled, albeit under circumstances that would have daunted a less fanatical aspirant.

Meanwhile on November 1, *Huáscar* had begun to earn her keep for her new masters, taking her former compatriot *Pilcomayo*. Henceforth *Union* was the sole seaworthy Peruvian unit, and *Pilcomayo* was incorporated forthwith into the Chilean Navy.

BLOCKADE OF THE PERUVIAN COAST
JANUARY – DECEMBER, 1880

The blockade of Callao was maintained by *Blanco Encalada*, *Huáscar*, *Abtao*, *Pilcomayo* and the armed merchant cruiser *Matias Cousino*. *Almirante, Cochrane, O'Higgins, Magallanes, Covadonga* and the armed merchant cruiser *Loa* functioned in the same capacity off Arica.

On April 22 and 25, the blockading squadron undertook the bombardment of Callao. Firing was at ranges of from 6000 to 7000 yards; the Peruvian forts replied, but little damage was done on either side. On May 10, the attack was renewed. A training ship and several barges were sunk in the harbor, and the docks were damaged.

It was on this occasion that *Atahualpa* got under way and engaged *Huáscar*. A shell from the 15-inch smoothbore Rodman guns of the monitor penetrated the side of *Huáscar* and flooded one of her compartments; the low silhouette of the monitor preserved her from the fire of her foe. *Huáscar* then attempted to ram, but another shell from *Atahualpa* struck her and disabled her vulnerable steering gear, and the monitor was able to withdraw once more into the harbor.

During the month of May, the Chilean Navy took delivery of four torpedo boats. *Fresia* and *Janegueo* were built by Yarrow, *Colocolo* and *Tucapel* by Thornycroft, both British shipbuilders. *Guacoldo* was a Herreschoff boat built for Peru, but seized in transit by the Chilean armed merchant cruiser *Amazones*. A sister boat made her way into Callao without incident, where the Peruvians christened her *Republica*. (The torpedo boat was at this time a recent development, and since it was both in vogue and relatively cheap, it was doubtless inevitable that Chile would invest in them: It must frankly be stated, however, that it is difficult to conceive of duty to which they were less suited than coastal blockade.)

In the predawn darkness of May 25, *Janegueo* and *Guacoldo* were patrolling near the entrance to the docks of Callao. The Peruvian steam launch *Independencia* (not to be confused with the broadside ironclad of the same name) was sighted at close range, and *Janegueo* at once bore in to attack. The crew of the Peruvian, although they lacked the McEvoy duplex outrigger torpedoes of *Janegueo*, nevertheless

51

were not behindhand in improvisation. As the torpedo exploded against her hull, the crew of *Independencia* hurled a 100-pound cask of powder onto the deck of *Janegueo*, and Captain Galvez detonated it with a pistol shot. Both boats sank simultaneously, the Peruvian with the loss of eight men out of her crew of 16. All the Chileans survived. This clearly indicates that they had already chosen to regard *Janegueo* as expendable. Indeed, they must have already been transferring to their lifeboat, which was towed astern, at the moment when the torpedo struck and *Independencia*'s unorthodox counterattack developed.

Having utterly dispersed the allied armies at the Battle of Tacna on May 26, the Chilean Army moved to surround Arica, which they accomplished June 5. On that date a surrender ultimatum was conveyed to the Peruvian commander, and the navy blockaded the port by way of emphasis. The ultimatum was rejected and the naval attack was answered by the Peruvian shore batteries and the guns of *Manco Capac*. The Chileans seem to have had the worst of the engagement; *Covadonga* was damaged, and *Almirante Cochrane* suffered 28 casualties when she was penetrated by a 15-inch shell from the monitor. On the other hand it is difficult to assess the damage caused by the Chilean fire, since the port was subsequently devastated by the Peruvians. By Jun 7 the Chileans were in control of Arica proper, but Captain Lagomarsino of *Manco Copao* succeeded in scuttling his ship, blowing up the shore batteries, and setting fire to the dock facilities before the enemy troops were able to intervene.

On July 3, an unidentified ship, having been sighted to the north of Callao, the armed merchant cruiser *Loa* was dispatched to investigate. The vessel proved to be an unarmed freighter, loaded with fresh provisions and abandoned at anchor. The prize was towed alongside *Loa* and the cargo transferred to her. As the last of the provisions was hoisted out of the freighter, a violent explosion tore open the sides of both ships. *Loa* sank in five minutes, the derelict a good deal faster. Thirty-eight Chileans survived out of a crew of 92. While modern sources invariably attribute this explosion to a submerged mine, an examination of the circumstances inclines me to believe that the initial Chilean interpretation was correct: the derelict itself was booby-trapped, with an explosive charge deep in the hold set to go off when the pressure of the cargo above it was removed. If this is in fact the case, it is a stratagem unique in the history of war at sea.

At about this same time *Covadonga* was detached to blockade the port of Chancay, twelve miles north of Callao. In the course of her duties, she sighted a steam launch inshore towing a gig astern. The launch was sunk by gunfire, and a boat sent to examine the gig. Finding it sound and serviceable, the boat's crew towed it back to *Covadonga*. As the gig was being hoisted aboard, *Covadonga* ran upon a submerged mine, the explosion of which stove in her starboard side. She sank rapidly, but the majority of her crew survived, although 49 of them were picked up by Peruvian boats from the shore and made prisoners. The Chileans at the time attributed this explosion to a booby-trap set in the gig, but it is clear from the circumstances that this was not the case.

On November 18, 1880, the Chilean Army disembarked at Pisco to attack Lima. They overran the first of two fortified lines before the city, but at such cost that they initiated peace negotiations. Two days later, with the conference already in session, Supreme Chief of the Republic Pierola observed an inadvertent maneuver of the Chileans that he believed left them vulnerable. He at once set in motion an assault upon their position, with the result that after four hours fighting the Peruvians were utterly defeated and dispersed, with Pierola dead on the field.

Negotiations were at once suspended, and the Chilean Army moved into Lima against negligible opposition. News of the defeat reached Callao before the Chilean troops could move down to occupy the port, and the remaining few ships of the Peruvian Navy were put to the torch. *Atahualpa*, *Union*, the captured transport *Rimac*, and a considerable fleet of small craft perished in the conflagration.

The army of Chile occupied Peru for almost three years before the war was formally ended by the Treaty of Ancón on October 20, 1883. Tarapaca province was ceded to Chile outright, with a ten-year Chilean occupation of Tacna and Arica to be followed by a plebiscite. Bolivia lost its access to the sea to Chile and became, and still is, a landlocked nation.

These conflicts are now largely forgotten, but they offer lessons that retain a value not only for the student of history, but for the designers of our future. They were observed closely at the time and the value of these lessons was not lost on their contemporaries. In many cases foreign observers were present during the battles.

NAVAL ACTION DURING THE CHILEAN CIVIL WAR OF 1891

The Chilean Civil War of 1891, also known as the Revolution of 1891, centered around a power struggle between the Executive Branch of the government, in this case President José Manuel Balmaceda and the Congress as to which should have the final authority to govern. Specifically, by custom in Chili at this time, the president could stay in office only if he had the support of the Congress and Chamber of Deputies. The specific maneuvers that ultimately led to open conflict do not concern us here. It is sufficient to say that the army supported the president while the navy supported the Congress. This was the first of three Chilean civil wars and coup d'état. It has been observed often that a nation that sees military action as the solution to external problems will often use armed force to address internal disagreements as well. The philosophy is difficult to contain within a single sphere of conduct.

While the navy as a whole sided with Congress, the government of President Balmaceda had possession of two newly purchased British-built torpedo gunboats, *Almirante Lynch* and *Almirante Condell.* The torpedo gunboat was an intermediate type that passed fairly quickly into the limbo of unsuccessful warships. Like "small elegant cruisers" in appearance, they were designed to be fast and well-armed enough to destroy torpedo boats before they could menace the capital ships of the fleet. As their speed ought to fit them for launching torpedo attacks as well, they were provided with a heavy torpedo armament, but they were quite simply too slow for their assigned tasks. It was not until the development of a reliable steam turbine that the necessary speed could be attained in a vessel of this size. By that time thinking and design had changed, and it was turbine-driven destroyers, rather than torpedo gunboats, which were completed in great numbers to escort the battle fleets. *Almirante Lynch* and *Almirante Condell* were completed in 1890 and carried 87 men. They had one-inch steel plate protecting the boilers and magazines; the conning tower was of one-inch steel as well. Two of the 3-inch (14-pounder) Hotchkiss QF guns were echeloned on the forecastle, with the third on the poop; one 14-inch torpedo tube was fixed in the bow, with two trainable tubes on each broadside.

The two gunboats were ordered out on the night of April 22-23 to cruise northwards toward Caldera and attack targets of opportunity. Finding nothing at sea, the two at 0400 on April 23, 1891 entered the harbor of Caldera. Within, they discerned *Blanco Encalada* at anchor near the shore. The gunboats approached to within 100 yards of the enemy unobserved, and at this range, *Almirante Condell* fired two torpedoes. These failed to score, so the two came yet closer, and at a bare 50 yards, *Condell* fired another torpedo, while *Lynch* fired two. The second of *Lynch*'s torpedoes struck the old ironclad amidships and detonated. The torpedo gunboats, their presence finally discovered, fled under a hot fire from the stricken vessel and from shore batteries; but they succeeded in escaping without serious damage. Within five minutes *Blanco Encalada* went to the bottom with 219 officers and men, the first warship ever sunk by a locomotive torpedo. Despite this noteworthy, albeit minor victory, the Congress ultimately prevailed.

The navy was active throughout supporting military operations and attacking such targets as were available. Of note was the protected cruiser *Esmeralda,* (not to be confused with the wooden vessel sunk at the Battle of Iquique in 1879). This ship was built to the design of George Wightwick Rendel for Chile by Armstrong in 1884. She was at the time the fastest warship afloat clocking 18.29 kt during her speed trials and is considered to be the first true protected cruiser. Her role in this conflict was restricted to bombardment of land forces loyal to President Balmaceda at Iquique and the critical Battle of Concon where her artillery led to the final collapse of the president's forces.

President Balmaceda committed suicide following this defeat, marking the transition from the Liberal Republic to the Parliamentary Era in Chilean history. *Esmeralda* was sold to Japan in 1894 on the eve of the Sino-Japanese war. Chile was treaty-bound to remain neutral at this time and therefore *Esmeralda* was sold to Ecuador and then resold immediately to Japan where she was renamed *Izuma* (sometimes spelled *Idzuma*), and served that country in the Russo-Japanese War participating in the Battle of Tsushima. She was among the first ships to make contact with the Russian Second Pacific Squadron.

We are now moving to a different navy and a new geographic area: the British in the Mediterranean Sea. We will note these differences, but it will be clear that Her Majesty's Navy was not without its flaws or its heroism. The time frame shifts backward a bit as well, but not so much as to interfere with the reading.

INTRODUCTION TO THE CAPITAL SHIPS AT ALEXANDRIA, 1882

We will concede at the outset that these are ugly ships; further, they seem to the modern eye rather wildly divergent from the subsequent course of warship evolution. The sailing rig seems anomalous, and the broadside gun ports appears to blindly follow the tradition of the old wooden warship, or at least to continue in that tradition for lack of any better idea. The concept of the turret, where it appears, seems to have been grievously misunderstood, or even willfully misapplied. The big muzzle-loading guns seem anachronistic: indeed, the Royal Navy, after a short experiment with breech-loading Armstrong guns, had in 1868, reverted to muzzle-loaders, a stance which was maintained until 1882, and which supplies modern critics of these ships with some of their most gleeful barbs.

A correct understanding of the limits of technology prior to 1880 allows us to view these ships in a different light. They were iron ships; the availability of cheap, uniform-quality steel had to await the development of Bessemer's blast furnace. Thus, the steam pressure within the boilers was limited by the strength of their material so that, as Archibald very rightly states:

> … the poor fuel economy of marine steam-engines made a … sailing rig an essential item for any ocean-going ship .

The average fuel supply for the boilers of vessels at this time was only 4 or 5 days at full speed. In addition, it must be kept in mind that the boiler might require hours to heat enough to provide full power so they were often run constantly at low speed. The screw could have considerable drag as well if it were not turning and while some vessels were equipped with a means of disengaging and hoisting it out of the water when not in use, this required additional time to re-engage it and therefore in practice the slow speed revolution was more common.

The length of an iron ship was likewise restricted by the limitations of the material. Anything over about 300 feet was not really feasible, for the ship became dangerously flexible. An elaborate system

58

of diagonal framing made possible lengths of up to 350 feet, but this was the absolute maximum. In a ship this short, the guns (or turrets), as the heaviest components, must be mounted quite close to the water. Any attempt to mount them on an upper deck would result in dangerous top-weight and the tendency to capsize (which was in fact the fate of *Captain*, a ship like *Monarch* but with the turrets set a deck higher, and which went to the bottom in September, 1870 with all but 17 of her company). Thus, the main armament was perforce mounted on the main deck, either in gun ports or in badly obstructed turrets.

As to the nature of the guns themselves, it was once again the relative weakness of iron that prohibited the breech-loader. The Armstrong BL guns in service prior to 1868 were involved in a number of disastrous accidental explosions. The fault was fixed upon the iron breech mechanisms, which were too weak to withstand the strain of repeated firing. To have waited until 1882 to reinstitute breech-loaders may seem a trifle extreme, but the disasters of the experimental ordnance had reinforced the Royal Navy's innate conservatism. (The philosophy of innovation was briefly this: British shipyards were the largest and most advanced in the world. Thus Britain could afford to abstain from experimental designs, allowing other countries to work out the problems of warship design in their own time and at their own expense. Then, a superior design of ship having been established by the trial and error of others, Britain could readily enough out-build the competition in the new types. It is also true that a great deal of the construction of other countries took place in British shipyards so the designs were well known to them.)

In short, then, these ships, for all their seeming eccentricity, were quite sensibly designed: they were the best that could be done with iron. Fortuitously, in 1882, the Mediterranean Fleet constituted a most representative sampling of the Black Fleet.

Inflexible was less than a year old in 1882. Conway calls her "a ship of extremes with the heaviest muzzle-loading guns in the Royal Navy, and the thickest armor ever put afloat." Her thickest armor was 24 inches, but only the central citadel was armored; this was compensated by the fact that the ratio of armored/non-armored hull space was such that she could remain afloat even with her unprotected ends waterlogged. "The 16-inch 80-ton guns fired a 1684 pound shell at the rate of one round per gun per minute." The muzzle-loading guns on *Inflexible* were depressed for reloading into a fixed armored chute leading into a working chamber; as they were too long to be withdrawn into the turrets. She carried 440 men. Her design was the work of Mr. Nathaniel Barnaby.

(For comparison the largest shell fired by a British vessel was the BL 18-inch Mk I, a breech-loading naval gun mounted on the *Lord Clive* class monitors firing a shell weighing 3,320-pounds. The Japanese *Yamato* class battleships of World War II were equipped with the largest guns mounted on any ship, the 40 cm/45 Type 94 naval gun, which were actually 46 cm and therefore slightly larger than their British competitor. One of the shells from the *Yamato* can be seen at Mikasa Park near Tokyo, Japan.)

Superb had been built for Turkey as *Hamidiah* to the design of Sir Edward Reed, but was purchased for the Royal Navy during the Russian war scare of 1878. She was designed for Mediterranean service and thus lacked an upper deck battery, which was normally provided so that it might be worked when heavy seas rendered the main-deck guns inoperable. Her thickest armor was twelve inches, and her crew was of 642 men.

Alexandra completed in 1877. She was the first Royal Navy ship fitted with vertical compound boilers, whose steam pressure was twice that of the older rectangular type; at the time of her completion she was the fastest battleship afloat. As a number of mastless battleships with turrets fore and aft (virtual prototypes of the classic pre-dreadnought battleship) were already in commission, it was, as Archibald states, "astonishing" that a masted central-battery ship should have been laid down, and *Alexandra* was indeed "obsolescent when new". She was Mr. Nathaniel Barnaby's design. Her thickest armor was twelve inches, and her crew numbered 674.

Sultan was designed by Sir Edward Reed, and completed in 1871. "The System of building out the upper-deck armored battery was taken a step further in *Sultan* in which it formed the widest part of the ship; it also, for the first time, had transverse armor, so that she had complete box-batteries on two decks." Her thickest armor was nine inches, her complement 655.

Temeraire, completed in 1877, designed by Mr. Nathaniel Barnaby, was "the last and the strangest of all the masted ironclads built for the Royal Navy." Two of her 11-inch guns were mounted on the upper deck, in pear-shaped barbette fore and aft; the upper deck of the central battery was omitted. The hydraulic turntables of the barbette guns, with the hydraulic transoms, that raised and lowered the guns were the design of Captain Scott. This arrangement gave her the greatest field of fire of any masted battleship. Her thickest armor was eleven inches, and her crew numbered 580.

"*Monarch* was the first seagoing turret ship and the first British warship to carry 12-inch guns." She was of Sir Edward Reed's design and she was completed in 1869. She was cut down to barque rig in 1872. Her thickest armor was ten inches, her complement 575.

Invincible was a central-battery ship of the *Audacious* class, the design of Sir Edward Reed completed in 1870. She and her three sisters were the largest class in the Black Fleet. They were rather small, intended as second-class ironclads for service on foreign stations. The central battery was on two decks; they were the earliest ships so provided. Their maximum armor thickness was eight inches, and they carried 450 men.

Penelope, completed in 1867 to the design of Captain Cowper Coles, was the first central-battery ship to join the broadside ironclads of the early Black Fleet. "The last of the small ironclads," Conway calls her, and she was in fact more a test-bed for the new central battery design than an Armored Corvette or any other standard classification. Her thickest armor was six inches, and she carried 550 men.

THE BOMBARDMENT OF ALEXANDRIA
JULY 11, 1882

The Khedive of Egypt, having mortgaged his country to the British and French in order to support his extravagant habits, found that the army under Arabi Pasha at length ceased to respect his authority. As provocations against European nationals increased, the British and French sent a combined squadron to Alexandria in order to maintain the peace. The arrival of the warships triggered rioting of a hitherto unprecedented violence. Moslem fanatics murdered 550 Europeans, including an officer and two men from the British ships; Arabi Pasha's army stood aside, and individual officers actively encouraged the disturbance.

Diplomatic activity, then in progress, prohibited an immediate reaction by the fleet, to the intense outrage and disgust of officers and men alike. In any event it was not until a month after the riots that the city was bombarded, and the pretext was not the death of the three navy men, but the continuing fortification of the city by Arabi Pasha's forces. The French at the last moment declined to become involved, and on the evening of July, 10 their ships steamed out of the harbor. *Alexandra*, the flagship of Admiral Sir Beauchamp Seymour, was saluted by each ship in turn as it passed, her band playing the French national anthem in response.

Firing commenced at 0700 with the signal "Attack the enemy's batteries". Despite the unnerving effect of the huge naval shells, the Egyptian gunners stood gallantly to their pieces, and the fighting continued much of the day.

There is really little enough else to be said concerning the action proper: shore bombardment is a static operation that in no way lends itself to narrative history. There are, however, a number of points to be made in response to a curious body of opinion that has grown up around this episode. There were at the time certain writers who, motivated by hostility towards Britain, distorted the events and removed from context certain superficially damaging aspects of the British reports. A school of modern historians has accepted this propaganda at face value, for which reason it requires refutation.

62

"In general, the gunnery of the British fleet was very indifferent. After the bombardment a close inspection of the forts showed them to be far from demolished. Almost all the guns might have been fought again." In fact, although indeed only three Egyptian guns were destroyed by direct shell hits, a great many more were buried by the earth thrown up by near misses or dismounted by the concussion of them. (It must be stated that a number of guns also seem to have dismounted themselves by their own recoil, due to their being so poorly emplaced.) That these guns "might have been fought again" is technically accurate; in order to work them, however, extensive digging out and repair would have been necessary, and would scarcely have been feasible in the face of the enemy. In the end, Egyptian fire slacked off and ceased, not because most of the guns were destroyed, but because ammunition ran out, and it was impossible to bring more forward to the forts under fire. It has always been the case with shore bombardment that relatively few guns are actually destroyed: the great majority of "knocked out" pieces in fact cease fire because their crews are driven to cover, or because they cannot be supplied under fire.

"Out of a total of 16,255 rounds fired from the Nordenfelts only seven found their mark." This is statistical manipulation of the most blatant sort. We are speaking of machine-gun fire, always the least accurate per bullet of any variety. We are speaking, furthermore, of Nordenfelt machine-gun fire. The Nordenfelt was, to speak charitably, not in the mainstream of machine gun evolution. It consisted of five or more barrels laid side by side atop a flat base, and fed by ammunition chutes perpendicular to the barrels. In appearance the whole contraption was not unlike a small hand press. It was operated by a manual crank at the right corner, and required to be jerked about quite vigorously by the gunner in the process of firing. It is therefore no surprise that, when the Nordenfelts were trained upon dust-shrouded fortifications which erupted with major-caliber shell explosion 2000 yards away, their accuracy was minimal. (In this situation, exactly when has a machine-gun round "found its mark", anyway? When it pings worthlessly against the barrel of an enemy artillery piece? When it strikes down a random gunner while those about him succumb to shell splinters instead? To speak in that manner is in truth not only misleading but ultimately virtually meaningless.)

"The chief lesson of that engagement was the disadvantage of black powder. The British warships had to interrupt their fire for quite long spells in order to let the smoke clear. This gave direct impetus to the search for smokeless explosives." The search for smokeless explosives was in fact already well under way. We have seen that gun- and coal-smoke were a serious hindrance to visibility in any naval action. Reflection will assure us that in this particular case, with the ships virtually stationary, and hence generating little coal-smoke, the problem must have been less severe than ordinarily. In fact, even if a ship had managed to surround herself with an impenetrable pall of powder-smoke, a few turns of the screws would have sufficed to move her clear of the cloud. I suspect, however, that this writer has misinterpreted his data. I would suggest that it was not powder-smoke around the ships that caused the obscuration, but the smoke and dust thrown up around the fortifications by shell bursts. Throughout the history of shore bombardment we find ships obliged "to interrupt their fire for quite long spells" in order to allow this latter variety of smoke to blow clear. The problem was finally alleviated, although by no means entirely solved, during World War II by the use of spotters.

"The *Inflexible* was in some ways an anachronism, being brig-rigged with two masts and enormous yards to carry sails. If felled in action they would not only have done deck-damage, but might probably have fouled the propellers. Yet they accompanied her into the bombardment of Alexandria, where mercifully the Egyptian gunners missed them." In fact *Inflexible* was hopeless under sail and rarely hung canvas; the rig was designed principally that the men might keep their hands in aloft, since the odds were that their next ship would be a genuine sailer. Further, we have seen that the sails were designed, on all masted ironclads, for cruising; the ships fought under bare poles, with the yards and sails stowed below. If *Inflexible* went full-rigged into action, this was not naval doctrine, but the individual aberration of her commander (as he was the eccentric Captain "Jackie" Fisher, it is not impossible).

In the course of the bombardment *Inflexible* fired 88 16-inch shells; 5052 shells of lesser caliber were expended, as were 55,500 bullets. *Inflexible* was also "the most heavily damaged of the British ships, being hit by several shells including one 10-inch. Some damage was also caused "by her own gun blast." (The latter is an extremely common phenomenon.) *Superb* "suffered minor damage". *Alexandra* "was hit 24 times but suffered no serious damage". *Penelope* "fired 231 projectiles, receiving only slight damage".

The squadron was prepared to resume action on the morning of Jul 12, but the Egyptian works were deserted. Arabi Pasha's forces had been withdrawn from the city during the night in anticipation of a British landing, which their commander judged them incompetent to resist. *Temeraire* and *Penelope* in fact stood in to the works during the morning and put ashore two boats whose men spiked such guns as seemed still operable, but no attempt was made to land a force in the city proper, which was now in the hands of the mob.

Admiral Seymour did not wish to risk the lives of his men in street fighting, and at any rate had no firm authorization from his government to land a force. For three days the mob ran rampant, looting and burning, until finally even they abandoned the city. On the morning of July 14, a group of about eighty Europeans who had taken refuge in the Anglo-Egyptian Bank building, out of food and desperately short of water, at last despaired of rescue from the fleet and determined to make their way to the waterfront. Setting out in a group, women and children in the center with the men around the edge, they made their way to the docks unmolested, picking up another twenty or so Europeans on the way. Commandeering small boats, they put out to the fleet and were taken aboard. The representations of these much-tried individuals at last convinced Admiral Seymour that there was no alternative to a landing in force, and on the morning of July 15, 800 men were put ashore with Gatling guns. Within two hours Alexandria was secured.

Arabi (Urabi) Pasha rose from the ranks of the Egyptian fellahim (peasant, farmer or laborer class) to the commander of the military and the power behind the government. He attempted to institute reforms particularly the elimination of Franco-British domination of the Khedive, but the British action at Alexandria ultimately brought about his downfall and the institution of a de facto British control of Egypt, the site of the all-important Suez Canal.

THE SINKING OF HMS *VICTORIA* JUNE 22, 1893

INTRODUCTION

This event is included for the information that can be obtained from its examination. It is presented here, somewhat out of its appropriate chronologic place, because the information is most useful in this portion of the overall presentation. Also there is a shift geographically from the Mediterranean to the Orient in the next section.

The ships involved will be only briefly described as opposed to the more detail descriptions usually provided because of the nature of this incident. It involved an accidental ramming and sinking of a British battleship by another British battleship and except for the details of the damage and results of the inquest it is of little real concern for the purpose of this text. The examination of the cause of the accident will be covered only briefly as well, but the examination of the damage and reasons for the rapid sinking of *Victoria* will provide some understanding of ramming as a weapon.

Victoria was commissioned in March, 1887, and at the time she was the largest battleship, with the largest guns afloat. Her armament consisted of two 16.25 inch, and one 10 inch breech-loading gun with 18 inch of maximum armor and a maximum speed of 16 knots. The *Camperdown* was completed in 1889 and carried four 13.5 inch breech-loading guns with maximum armor of 18 inches and had a maximum speed of 17.1 knots

The British navy had deified Admiral Horatio Nelson, but had. by this time, lost the lessons taught by him. Specifically the initiative of its officers had been stifled by the long peace that had followed Trafalgar. The Mediterranean fleet in which these two ships were serving along with 6 other battleships and 3 cruisers was the pride of British naval prowess since, after the home island, defense of the Mediterranean was deemed the most important responsibility of Her Majesty's Navy.

Vice Admiral George Tryon commanded this fleet and he was attempting to imbue his command with that sense of initiative it had lost. As a result, his orders were often obtuse and he apparently had a

66

habit of altering them at the last minute. His officers came to expect this.

THE COLLISION

On the evening of June 22, 1893, near the port of Tripoli, Lebanon, Admiral Tryon had his fleet in two columns, line ahead, steaming on parallel courses. He was on *Victoria* leading one of these, with Rear Admiral Markham on *Camperdown* leading the other. Tryon ordered the two columns to turn toward each other in order to reverse course and anchor for the night. Several officers pointed out that the columns were only 1200 yards apart and that more room was needed to accomplish the proposed maneuver safely. The turning radius of the ships was 800 yards making a collision very likely. While Tryon seemed to acknowledge this, he made no change in the order, indeed when Markham on *Camperdown* hesitated, Tryon signaled asking why there was a delay. The signal was sent by flag and therefore was visible to the entire fleet stinging Markham to action. He had considered semaphoring to Tryon his misgivings, but abandoned this idea. Thus, the columns turned and while many officers would later say they feared a collision, none questioned or took any action to avoid it until it was too late. *Camperdown* rammed *Victoria* forward of her belt armor and while at first it appeared she might reach shore before she sank, it was not to be. Three hundred fifty-nine men perished with her including the one man who might have explained the cause of the catastrophe: Admiral Tryon. Eight minutes after the collision the forward end of the ship was flooded and water was pouring in through the gun ports and deck hatches. *Victoria* capsized just 13 minutes after being rammed and sank shortly thereafter, her screws raised out of the water still turning.

THE INQUEST

The sinking was tragic and high profile and therefore demanded an explanation and the fixing of blame. The most interesting aspect of the subsequent inquest for the sake of this work however is the insight into the nature of the ram as a weapon. The "trial" was carried out in the naval court and in the press. Tryon, Markham and others, even the governing rules of the Royal Navy were condemned and acquitted in turn. The ultimate responsibility seems to have been shared by many. A single quote from the courts martial findings will illustrate:

> 'The court strongly feel that although it is much to be regretted that Rear Admiral Albert H. Markham did not carry out his first intention of semaphoring to the Commander-in-Chief his doubts as to the signal, *it would be fatal to the interests of the service* to say that he was to blame for carrying out the directions of the Commander-in-Chief present in person. (My italics).

This statement, taken from Gordon, *Rules of the Game*, clearly fixed a degree of blame on Admiral Markham, but refrained from actually accusing him in the interest of preserving the image of the Royal Navy.

Some points are worth looking at more closely however. *Victoria* sank quickly contributing greatly to the loss of life. This was finally determined to be the result of the fact that the bulkhead doors were not closed in time to prevent a rapid flooding of the forward compartments. Normally it required 4 to 5 minutes to accomplish this task. Attempts were made, but were impossible with the flow of water that was present. Indeed the inquest concluded that the limiting feature of the initial flooding was the rate at which water could flow through these open doorways not the hole in her hull. None was closed at the time of the collision and the crew was not at battle stations to close them when needed after the collision. Men were dispatched to accomplish this but were unable to do so before the sea began flowing over the bow and through the hatches and gun ports on the forward deck. If the ship's crew had been at battle stations *Victoria* might not have sunk at all or at least not so rapidly. It was of course not expected that a simple maneuver to anchor ships for the night would require battle stations, but it is interesting to speculate that the sinking of *Victoria* might not have taken place in a battle situation even if the vessel had been rammed. It is also noteworthy that *Camperdown* was seriously damaged as well, and was considered at one point to be in danger of sinking. Her ram was nearly wrenched off and she took on tons of water.

No flaw in design was given serious consideration during the inquest and specifically it was not judged that automatic bulkhead doors would have saved the ship. Despite this conclusion, automatic bulkhead doors were installed throughout the Royal Navy shortly after this accident. Water tight compartments were common practice at that time, but the inquiry denied that their specific placement on *Victoria* contributed to her rapid sinking. Again in-spite-of this denial, the bulkheads were modified on the sister ship, *Sans Periel*, and redesigned on future construction. The ram had passed from its prime position in ship design by this time in any event.

The conduct of various officers was criticized of course, but no change in the orders of conduct was effected. The duty of junior officers (Admiral Markham in particular) to obey his commander's order was cited as was the conflicting duty of a subordinate to disobey an order that obviously place his ship in jeopardy. That these two were contradictory seemed not to be overly bothersome to the inquest. As for Tryon, his unorthodox practices were cited, but not openly criticized. Such blame would have begged the question as to why he was allowed to pursue this practice. In truth, the Royal Navy itself was under the spotlight and it was acquitted by its own judgement.

There are some interesting side issues surrounding Admiral Tryon and the sinking of the *Victoria* however. One is that he had apparently complained to several people that he had recently had brief episodes of confusion. This might have been transient neurologic ischemia (transient ischemic attacks or TIA's) and may have interfered significantly with his judgement and decisions on the evening of the collision.

70

Also of note is that John Jellicoe was an officer on *Victoria* and survived to become commander of the Grand Fleet at the Battle of Jutland, the largest encounter of battleships ever to take place.

One final piece of interesting history: there is the persistent story that at the time of the sinking of *Victoria*, several people "saw" Admiral Tryon at a party in his English home two thousand miles away. The party was hosted by his wife who was unaware that she had already become a widow. Several attendees at this party, there were over two hundred, claim to have seen the admiral in full uniform walk down the central stair across the banquet hall and then down into the basement of his home. This is discussed at length by Edward Rowe Snow, the great chronicler of maritime lore and a favorite of mine, in his book: *Tales of Terror and Tragedy*. I do not usually including such references in a book devoted to factual information as is this volume, but this is such a tantalizing story and Snow is such an entertaining writer that I could not resist. The reading of his books in my youth is responsible in no small part for my interest in naval history and the pursuit of this volume.

Again there is a shift in geography, this time to the Asian waters with a brief inclusion of Cuba, in the Caribbean. There is also a bit of time adjustment, but this is compensated for by allowing the reader to make use of the maps provided in Appendix C page 266 for the Sino-Japanese and Russo-Japanese Wars.

SINO-FRENCH WAR, 1884-1885

INTRODUCTION TO THE SHIPS OF THE SINO-FRENCH WAR

The French cruisers were all three-masted vessels with the general configuration of sailing ships. *Bayard* had a prominent bridge structure abaft the foremast and two small funnels close together amidships with a single protective trunk enclosing them both and rising to about half their height. Two of the 9.4-inch/19 guns were in shielded sponsons on either side of the bridge. The other two were mounted in similar shields on the centerline, one amidships, one aft. One of the 7.6-inch guns was mounted under the forecastle and fired ahead, the other beneath the poop firing aft. The six 5.5-inch were in broadside battery on the main deck. Her thickest armor was 10 inches, and her complement was 451 officers and men.

Triomphante had a prominent forecastle and a single funnel that diminished in size about two-thirds of the way to the top. Two of the 9.4-inch/19s were in conspicuous barbettes with domed hoods on either side of the funnel; the other four were located on the main deck of the battery. The single 7.6-inch was a bow chaser, mounted beneath the forecastle; the six 5.5-inch were positioned on the upper deck. Her armor was up to 6 inches thick, and she carried a crew of 582.

Duguay-Trouin had two small raked funnels behind a rather inconspicuous bridge. Four of the 7.6-inch guns were sponsoned on either side fore and aft without protection; the fifth fired ahead. One of the 5.5-inch fired astern, the other four being mounted in broadside. This vessel was distinguished by a particularly prominent ram bow. The crew was made up of 522 men.

Villars and *d'Estaing* were almost identical in appearance, having a single good-sized funnel, with the bridge not at all conspicuous. Two of the 5.5-inch guns were mounted on a raised platform on the forecastle, the rest on the upper deck broadside in slightly sponsoned ports with thin iron shields. Both ships carried crews of 264.

The appearance of the lesser vessels is impossible to ascertain with any accuracy. We must note that at this time the term torpedo did

72

not connote the self-propelled "fish" or Whitehead locomotive torpedo that we now mean by the word. That device was still in an embryonic state in 1885, and none was available in the South China Sea. Torpedo still indicated what we today would call a mine, an underwater charge detonated either by impact, or electrically from the shore, and originally planted in shallow water. The effectiveness of underwater explosion in the destruction of warships, as evidenced during the United States Civil War, made it obvious that a means of delivering an underwater charge in deep water was most desirable. Two methods, both so crude as to be suicidal in the face of an even moderately resolute or alerted enemy, were devised. The "spar" or "outrigger" was a torpedo charge attached to the end of a long pole extending over the bow of a small steamboat which would run up to the enemy vessel detonate the charge against its hull, and then (hopefully) back clear. The "towed" torpedo was a floating mine pulled behind a similar boat (or even the same boat) at the end of a long cable, and maneuvered in a manner not dissimilar to that employed in towing a water-skier. The intention was that the boat should steer across the bows of the target; thus in due course as the enemy advanced the mine would be drawn into contact with his hull and would explode. The overpowering difficulty with both of these systems was that the enemy warship was scarcely likely to maintain course and hold its fire: it was far more apt to ram or gun down the torpedo boat, either of which a major warship could accomplish easily enough.

The larger Chinese vessels were all three-masted steamers on the model of a sailing ship. It is pointless to speak of them individually, for no reliable representations of them have survived. The names of the various Chinese vessels have differing English spelling since these are basically phonetic mimics of their Chinese pronunciation. I have chosen the most commonly used English names here, but the reader should be aware that others exist.

We must treat at some length the subject of Mr. George Rendel's "flatiron" gunboats, since they are of a type which was once quite numerous, both in the Royal Navy and in the navies of Armstrong's many client nations and the affair in the Pagoda Roads of Foochow was the sole action that any of them saw. They were shallow draft vessels mounting a single limited-traverse, forward-firing 10-inch muzzle-loading rifle forward in an armored box. In theory, they might charitably be described as self-propelled shore batteries. It was intended that, upon the appearance of the enemy, the gunboats should issue from the threatened port, proceed to appropriate points along the coast, and back themselves onto the shore (for which purpose they were provided with an auxiliary rudder in the bow), and engage the enemy upon his approach. It was reasoned that as they would be already aground at the outset of action the enemy could not sink them, but would have to pound them to pieces with gunfire, a much more laborious and time-consuming undertaking, during the course of which the Rendel boats might be expected to inflict significant damage upon him. They were totally unseaworthy, and since they were not meant to maneuver during battle, their engines were underpowered. The major weakness of the Rendel boats was the same as that of the primitive torpedoes: the assumption that the enemy would remain inactive. It is unreasonable to believe that any fleet, already in visual contact with the enemy port, would permit flotillas of gunboats to deploy along the coast and beach themselves without taking preventive action, and of course Rendel boats, if attacked while under way, were utterly defenseless. Although it is difficult to imagine that any sensible person would fail to recognize this appalling deficiency, in fact the Rendel boats were a popular item. We may venture to suspect that this is because they seemed to offer a new route to that El Dorado of appropriations committees everywhere: war on the cheap.

As an interesting aside, the United States Navy entertained a similar expediency in its early years under the presidency of Thomas Jefferson termed the "gunboat" navy. The same arguments favored this fleet of small, cheap vessels that were suitable only for harbor and coastal defense as their unseaworthiness made anything else virtually impossible. They were soon abandoned and while the navy processed about 170 gunboats at the beginning of The War of 1812, fought against Great Britain, none saw any serious service and the glory for the United States went to her frigates particularly the USS *Constitution*. It is little wonder that this was the case since one frigate had the gun power of 30 to 40 gunboats and the real cost of both construction and maintenance was approximately the same for a seaworthy frigate and the equivalent number of unseaworthy gunboats. Parsimony in governments often dictates acquisitions, however faulty that policy may be. It is still true, even to this day and I am afraid it will continue.

NAVAL ENCOUNTERS OF THE SINO-FRENCH WAR

(Since this conflict is now virtually forgotten in the English-speaking world, I feel obliged to enter briefly into matters which are beyond the scope of this present work. The Sino-French War, or "Second Tonkinese War" as it is called in French sources, was the result of French efforts to extend their Indo-Chinese protectorate into Tonkin province. The Chinese resisted, and for the last time the Imperial Chinese Army won victories in the field. Honors were fairly evenly divided on land, with perhaps a slight edge to the Chinese; the Treaty of Tientsin, Jun 9 1885, restored the status quo ante bellum.)

Although French foreign policy vacillated during this war, for France was in the throes of one of those peculiarly virulent Republics that she brings forth at intervals, the policy of the French Navy was simple and unswerving. Its objective was the destruction of Chinese naval power in the South China Sea, and this was in the main achieved, hobbled though the navy was by ambiguous and contradictory directives from the Paris government.

Following a French defeat on land, Admiral Courbet was ordered to establish a close blockade of the Chinese squadron at Foochow (also spelled Fozhou). Foochow is located well inland along the River Min; the entrance channel from the river mouth is rather narrow, but below the city the river widens out into the Pagoda Roads. In this roads the Chinese squadron was anchored; here also was the Arsenal, located immediately ashore.

The Chinese ships were under the command of Admiral Ting Juchang, an officer of mediocre achievement whose melancholy fate it was to preside, during the ensuing twelve years, over the virtual annihilation of his country's navy. In the inlet of the Min leading to Foochow proper were positioned: *Yang Wu,* Ting's flagship, *Fu Po* and *Fu Hsing* anchored in such a way that their guns and those of the Arsenal might be mutually supporting, and that access to the city was barred. Across the river, about two-thirds of the way from Pagoda Island to the far shore, *Chi An, Fei Yuan* and *Chen Wei* rode in column at anchor. There were a number of armed junks about, and the ridiculous Rendel gunboats lurked still further upriver.

76

Admiral Courbet chose to interpret his orders in the most rigorous fashion; he established a "close blockade" which must have seemed certain to exasperate the Chinese into opening fire upon him without much delay. Judging that *Bayard* drew too much water to enter the roadstead in safety, he assigned that ship to keep watch by the river mouth, shifted his flag to *Volta*, and led the remainder of his squadron into the river. Entering the Pagoda Roads without opposition, the French warships came to anchor. *Volta*, *Lynx*, *Vipere* and *Aspic*, in that order, anchored in left echelon across the mouth of the inlet, about 400 yards from *Yang Wu*, *Fu Po* and *Fu Hsing*, and somewhat nearer to the Arsenal. In the river proper, *Duguay-Trouin* squared off against *Chi An*, *Villars* against *Fei Yuan*, and *d'Estaing* against *Chen Wei*. The French ships thus formed a shallow V, or better, an inverted A with *Triomphante* along the crossbar, while the Chinese ships were divided by the French into two separate groupings. This was capped by an insolent message to Admiral Ting, advising him that his ships were under close blockade, and that any attempt to get under way would be regarded as provocative.

Supinely imperturbable, Ting responded that he must apply to Peking for instructions whereupon the fuming Courbet was obliged to apply likewise to Paris, for he had no authorization to attack an unresisting Chinese force. Then, while the matter wended its way through the tortuous intrigues of the Imperial Court and the no less Byzantine offices of the Third Republic, the two fleets rode at anchor, faced off but silent, for a full six weeks.

The French plotted every facet of the destruction of their erstwhile foe: each gun was sighted in, drills were conducted time and again, the precise moment at which the tide would begin to ebb and swing the sterns of the Chinese warships toward the French broadsides could be predicted by the color and appearance of the river. The Chinese, with truly Oriental stoicism, awaited their fate with indifference and made no visible preparation for the impending assault.

At length, on August 22, 1884, the long-desired authorization arrived. A flurry of signal flags aboard *Volta* shortly summoned the French captains to confer with their chief. It was ordered that, the tide having that day already passed its peak, fire should be opened on the following day at 1400 that being the hour of the tide.

In order to eliminate Admiral Ting at the outset, it had been decided that prior to the gunfire attack *T46* should run a torpedo against his flagship (There is some disagreement as to whether a spar or a towed torpedo was used, but on balance the spar torpedo seems indicated.) Precisely at 1400 the little boat began the 400-yard run-in. The Chinese made no response, and the attack went off like an exercise. The 30-pound charge detonated against the side of *Yang Wu*'s hull below the water, the blast threatening to swamp *T46* momentarily, but the boat backed clear undamaged. Immediately the French ships opened fire; the first shot seems to have been *Lynx*'s.

Volta took *Yang Wu* under fire as the holed sloop made for shore, and the Chinaman quickly caught fire, her crew going over the side. *Fu Po* and the three vessels in the river proper offered no resistance whatever, their crews able to think of nothing more than abandoning ship as soon as the French opened fire. The more impetuous dove into the water while the steadier hands endeavored to launch the ships' boats as a means of escape. The one-pounder revolvers (a sort of oversized Gatling gun firing explosive shells) on the French ships took the fugitives under fire, smashing the boats and working great execution.

The armed junks made a better showing, opening a hot fire, but their frail hulls and rice paper sails were shattered and set ablaze by shellfire and flames spread from one to the next. In short order they too were abandoned and afire.

Only *Fu Hsing* got under way as firing commenced, and *T45* at once made for her. Passing directly under the bows of the Chinese gunboat, the torpedo boat succeeded in drawing a floating charge against the enemy's side, which it detonated and blew a hole into the hull at the waterline. The towline briefly fouled *Fu Hsing*'s screw, and before it pulled free the immobilized *T45* was subjected to a heavy but inaccurate fire from her victim and from the Arsenal. The torpedo boat escaped without significant damage, and *Fu Hsing* ran for shore, under fire and burning. It was 1407 and only seven minutes into the battle.

Having just witnessed the first and only use of the towed torpedo in action, the Pagoda Roads was now to be the scene of the entire combat career of the Rendel gunboat. *Chein Shang* and *Fu Shang*, advancing far enough down river to bring the French ships within range, then opened fire and began backing, holding themselves steady with both rudders, meaning to stand in one place as their engines neutralized the current and bombard the French warships. The situation was momentarily awkward for the French, for their major vessels were still at anchor, and few of their guns bore upriver. The Chinese fire was not particularly accurate, however, and it quickly became obvious that the gunboats were not powerful enough to hold their own against the combined river and tidal current. In relatively short order they were borne down past the three cruisers in the river proper, which shattered them with a few broadsides, and they sank.

Yang Wu had already gone down, the anchored Chinese ships were hopelessly ablaze, but *Fu Hsing* was still under way. A boarding party being dispatched in launches, the Chinese crew was driven over the side without difficulty; but the gunboat was too far gone in flames, and the French had barely time to regain their launches before she went down, "sizzling and sputtering." Chinese casualties amounted to 521 men killed and 150 wounded, along with several hundred missing (but did not include Admiral Ting Juchang, who escaped ashore and whom we shall encounter again). Courbet gave his own losses as 6 killed and 27 wounded.

"La Grande Gloire du Foochow", (The Great Glory of Foochow), this affair is called among the French, in the annals of whose navy it occupies an exaggerated position, analogous to the Battle of Manila Bay in US naval histories. It was rather acidly labeled "The burning of the war junks," fifteen years after the event.

During the remainder of the afternoon Courbet's fleet silenced the shore batteries emplaced about the Pagoda Roads, and the following day was devoted to the reduction of the Arsenal. Proceeding down the Min, the French then engaged the forts at the mouth thereof; although they were taken in the rear, the destruction of the works continued from August, 25 to 29.

During the succeeding months, the fleet was occupied in extensive operations against Formosa, as well as the occupation of Mekong and the Pescadores, the detailed history of which does not properly concern us here.

The only further event of the struggle which we must consider is the destruction, on the night of February 14-15, 1885, of *Yu Yuen* and *Teng Ching*. These last two significant Chinese naval units in the South China Sea had been tracked down at Sheipoo, and two spar torpedo boats from *Bayard* attacked them under cover of darkness. *Yu Yang* was successfully holed by the charges, and *Teng Ching* was sunk in error by the wild firing of her stricken consort. *Yu Yang* was also a total loss, although the assertion that she exploded seems false.

THE SINO-JAPANESE WAR, 1894-1895
(It may be helpful to consult the map in Appendix C, page 266)

INTRODUCTION TO THE SHIPS OF THE SINO-JAPANESE WAR

A few general points must be made before we discuss individual warships. These vessels were designed during a period of rapid and fundamental transition. The traditional configuration of the three-masted warship with guns mounted in ports or embrasures along the side of the hull had finally been abandoned as unsuited to the requirements of newly-developed weapons and equipment. Naval architects, slowly and not without miss-steps, were formulating the principles and deducing the practices that would enable them to employ the new weapons to greatest advantage. Each new warship design was in a real sense experimental, and concepts developed so rapidly (as indeed did new inventions in engineering and weaponry) that a ship might easily be obsolete before she was commissioned.

The ideal battle fleet is a balanced, coordinated force, with each unit having much the same firepower and performance as every other. It will be obvious, however, that under the circumstances described above, this ideal could in nowise be approached. Even the elite Royal Navy was at this period characterized quite accurately as "a collection of samples." This lack of uniformity is greatly magnified in the present case by the fact that both Japan and China were third-rate naval powers, their fleets motley grab bags of warships purchased from foreign shipyards, or produced at home to foreign designs. There was not even that underlying coherence of design, found in the British or French navies, which was simply the continuity of a staff of naval architects and an integrated production plant.

The huge-gun craze of the 1870s had largely abated (although many fighting ships were still considerably over-gunned), and conventional gun-foundry was involved in the systematic improvement of old-pattern rifled breech-loaders of between ten and thirteen inch caliber. On the other hand, the new 4.7-inch and 6-inch quick-firer

(QF) were heralded by many as harbingers of the decline of the big gun.

The basic old-pattern breech-loader had a screw breech and used bagged charges, whereas the quick-firer had a sliding breech-block and employed a one-piece brass cartridge like an oversize bullet. The old pattern breech-loader had a hollow screw through which the shell and charge were inserted, and a heavy block was place between the breech and screw and then tightened. The process was long and the seal, and therefore the tightness of the breech closure, was dependent on the force applied to the screw. Several advances made by Thorsten Nordenfelt, William Armstrong and at Krupp Industries solved the problem with the interrupted screw mechanism or sliding breech which allowed for a much faster and much stronger breech closure. The breech was closed with a single one-third turn.

The real discrepancy between the respective rates of fire of the quick-firers and the main-battery gun did not stem from this difference, however. The fact was that barbette-mounted old-pattern guns had to be lowered to 0° elevation and trained along the centerline in order to reload, since the sub-turret working chamber for all-around loading had not yet been developed. The quick-firers had a hydraulic recuperator and was trunnioned to a rotating pedestal-mount, thus facilitating continuous reloading through a considerable arc. The sum of these considerations is that while the 12.6-inch Canet rifles of the *Hashidate* class fired one round in five minutes, the 4.7-inch QF fired twelve aimed rounds per minute.

It was not disputed that a 4.7-inch QF gun was vastly superior to an old-pattern 4.7-inch. The weakness of the quick-firer concept was the more massive construction and finer tolerance demanded by the sliding-block breech as compared to the older screw breech. To construct a quick-firer of any caliber greater than 6 inch was beyond the skill of metallurgists and gun-founders at this time.

The partisans of the big gun maintained that a 6-inch shell lacked the momentum and bursting charge to significantly damage a major warship. It was their contention that however useful the quick-firer might be for anti-torpedo-boat work, the larger guns ("ship-smashers," they liked to call them) remained indispensable for fleet engagements.

The quick-firer's adherents regarded the big gun as obsolete. It was their belief that, while a single 4.7- or 6-inch shell might have no chance of crippling a fleet unit, the cumulative effect of ten minutes' continuous fire from the dozen or so quick-firers of a modern cruiser would disable any old-style battleship before her ponderous rifles had time to range in. ("A rain of fire" was their catch-phrase.) It is possible to discern the influence of one or the other of these opinions in the construction of the individual warships. How they proved out in action we shall see.

The locomotive torpedo of 1894, steam-powered and discharged from a tube by compressed air, carried a warhead of about 300 pounds, a distance of 1000 yards at 30 knots. The torpedo boat was still very much alive; it had grown since the 1880s and in addition to its torpedo tubes on swivel mounts above decks, it carried three or four small QF guns (so that it might attack enemy torpedo boats in defense of the fleet). Most major warships also carried torpedo tubes. Almost invariably there were four, one fixed in the bow, one likewise in the stern, and two trainable tubes, one on either side amidships.

A few words concerning the practice of rating cruisers as "protected" or "armored". The protected cruiser was distinguished by a convex protective armor deck, which joined the hull sides below the waterline and curved upward to shield the magazines and vital machinery. It was pierced by funnels and ventilators, and sometimes an armored barbette joined it and continued upward to the upper deck. It was reasoned that shells might be allowed to pass through the superstructure with relative impunity.

It must be recognized that the protected cruiser and the armored cruiser were not contemporaries: the concept of the protected cruiser was at length judged wanting, and armored cruisers then began to be built. These were distinguished by an armor belt along the sides of the hull, with sometimes a lighter armor upper deck. The belt did not extend all the way to the upper deck: in cases of inept design, such as the *King Yuan* class, it was entirely below water at combat load, and thus of dubious advantage.

The intention was generally that the belt should extend to six to ten feet above the waterline. The barbettes were lightly armored behind the belt, and then heavily protected as they passed through the unarmored upper hull. The ships with which we now deal were for the most part designed before either of these concepts had been fully articulated, and were originally classed simply as cruisers. Later, the more specific designations were retroactively applied to them (not without anomaly--*Chiyoda*, for example, scarcely fits the idea of the armored cruiser as it is generally understood, but she is rated thus for her relatively thick plate). I have nonetheless followed the rating system as it is applied in Conway, chiefly in the interest of coherence.

The Chinese fleet was trained and advised by foreign officers (among them an American, Philo McGiffin, the Annapolis-educated captain of *Chen Yuen*) an advantage which the Japanese did not share. It was anticipated particularly by Western observers, that this would give the Chinese a considerable edge.

CHINESE VESSELS

Ting Yuen and *Chen Yuen*, the heart of the Beiyang Fleet, were German-built battleships, already ten years old in 1894. Their guns were Krupp breech-loading rifles of the old-pattern. The hooded barbettes were arranged in echelon, *Ting Yuen* having the forward barbette to port, *Chen Yuen* to starboard. The 5.9-inch were mounted at the extreme bow and stern. The maximum armor thickness was 14 inches, and they carried 550 men.

King Yuen and *Lai Yuen* were also German-built, six years old. Their guns were Krupp old-pattern breech-loaders. The two 8.2-inch were mounted in a twin barbette forward and the two 5.9-inch in single sponsons amidships. Their thickest armor was 9.4 inches, their complement 270.

Ping Yuen requires a little elaboration. She was the first Chinese-built armored ship, laid down in 1885 and designed as a battleship of the dimensions of *Ting Yuen*. Over the years Chinese incompetence made necessary the revision downward of her dimensions and armament, and even so in the end Chinese gun-founders proved unable to arm her, and her breech-loading old-pattern rifles were purchased from Krupp. Completed as a battleship in 1889, she was later retro-rated an armored cruiser. In fact, her extremely low speed rendered this designation inappropriate as well, and the only role in which she could serve was that of coastal gunboat. The 10.2-inch was in barbette forward, with the 5.9-inch in sponsons amidships. Her thickest armor was 9.4 inches, and she carried 202 men.

Chih Yuen and *Ching Yuen* were seven-year-old British-built protected cruisers. Their guns were Armstrong breech-loading rifles of the old-pattern. The 8.2-inch were behind open shields, twin mount forward and single mount aft; the 6-inch in sponsons amidships. The 57-mm (6-pounder) QF guns were anti-torpedo-boat weapons, too small to be effective against major warships. The thickest armor was four inches (gun shields two inches) and the crews were of 260 men. *Chi Yuen* was German-built, a protected cruiser nine years old. Her guns were old-pattern Krupp rifles: the 8.2-inch were mounted in a hooded barbette forward, with the 5.9-inch aft. The 3-inch were amidships. Her barbette armor was fourteen inches thick, although the protective deck was only a maximum of four inches and her complement was 200.

Chae Yung and *Yang Wei* were thirteen-year-old British-built protected cruisers. The 10-inch Armstrong breech-loading old-pattern rifles were mounted on revolving turntables in fixed gun houses fore and aft, with the old-pattern 4.7-inch arranged at each corner of the superstructure. Their thickest armor was only one inch, and they carried 177 men.

Unfortunately it is impossible to determine the appearance of *Kuang Chia*. She was a Chinese-built composite-hulled dispatch vessel, six years old. Her guns were Krupp old-pattern rifles; their arrangement aboard is unknown. She was unarmored and carried in the neighborhood of 150 men.

Kuang Yi and *Kuang Ping* were Chinese-built gunboats, only a year or two old. Their three Krupp 4.7-inch QF guns were mounted: one on each side of the bridge and one aft. The conning tower had two-inch armor and the crew was of 120 men.

Tsao Chiang was a domestically produced composite-hulled gunboat provided with a full spread of sail. She was around fifteen years old; the particulars of her 6.5 inch guns are unobtainable. She was unarmored and shipped a crew of 79 men.

Fu Lung (S10) was a standard German-built first-class torpedo boat, eight years old. One 14-inch Schwarzkopt torpedo tube was fixed in the bow, the other on a trainable mount aft; one reload per tube was carried. An unarmored vessel, she was manned by a crew of twenty. The subject of other Chinese first-, second~ and third-class torpedo boats is one of the more obscure minor matters of this period. The Chinese appear to have kept them lying about the harbor, so to speak, without even attempting to name or number them, or to keep any record of their individual activities. Such log-books as they may have had were lost. We have the records of their German builders concerning their specifications and dates of delivery, and we have Japanese records of the three they eventually captured; in the interim their history is blank. Under these circumstances, and since none of them engaged successfully, I shall omit to list their individual specifications, which would be meaningless. They ranged from 15.7 to 64 tons displacement, were between 64' 9" and 114' 10" in length, made between 15 and 18.5 knots, carried one or two 1-pounder QF guns and one or two 14-inch torpedo tubes, were between ten and thirteen years old, unarmored, carrying sixteen to twenty men.

JAPANESE VESSELS

Chiyoda was British-built, four years old in 1894. It had originally been intended that she carry two 12.6-inch Canet rifles, but the excessive topweight anticipated from this, together with a faith in the new quick-firers, caused the Japanese to propose a change to ten Elswick 4.7-inch QF guns. These were mounted eight in broadside sponsons and one each as bow and stern chasers. Her armor belt was 4.5 inches, her complement 350.

Of the *Hashidate* class Conway observes drily that "the design was not a success." They were massively over-gunned, and protection everywhere but around the main gun was absurdly inadequate. I may also say that they were among the most unsightly warships ever completed. *Itsukushima* and *Matsushima* were of French manufacture while *Hashidate* was made in Japan to the French designs. The French ships were three years old, while *Hashidate* completed just in time for the war.

Itsukushima and *Hashidate* mounted the old-pattern l2.6-inch Canet breech-loading rifle forward, while in *Matsushima* it was aft; the 4.7-inch QF were in broadside battery (with one as a stern chaser in *Itsukushima* and *Hashidate*). The maximum thickness of the protective deck was two inches, although the barbette was twelve-inch-thick armor. The ships carried 560 men.

Yoshino was arguably the finest ship in the Japanese fleet. She was British-built and barely a year old at the battle of the Yalu River. Two of the 6-inch Elswick QF guns were sited fore and aft with the other two sponsoned forward on either side. The 4.7-inch QF were in beam sponsons. Her protective deck and her gun shields were of 4.5--inch armor, her complement 560.

Naniwa and *Takachiho*, British-built cruisers eight years old, were balanced ships, although a trifle over-gunned. The 10.5-inch Krupp old-pattern rifles were barbette-mounted fore and aft; they probably still mounted their original old-pattern 5.9-inch (these were exchanged for QF guns "about 1895"), in battery amidships. The thickest armor was three inches, and the crew was of 325 men.

Akitsushima was built at Yokosuka from imported materials to a British design. She was completed in 1894. Jentschura, Jung and

Michel assert that originally it was intended to arm her with a 12.6-inch Canet, but this is certainly false. She was in fact a reduced copy of the United States cruiser *Baltimore*, which like her, was designed by Sir William White. Two of the 4.7-inch QF guns were mounted as bow and stern chasers, while the others, as well as the four 6-inch QF, were in broadside sponsons. Her gun shields were 4.5 inches thick (deck three inches) and her crew numbered 550.

Fuso was a sixteen-year-old British built ironclad of totally obsolete pattern. Her four 6-inch QF guns had been added recently, at the same time her rig had been cut down from its former three-masted configuration. Her thickest armor was nine inches, and her complement was 250.

Hiei was British-built, as old as *Fuso* and of a design a little more antiquated. None of her broadside-mounted guns was quick firing, and she still carried her full barque rig (although we may assume that she fought under bare poles). Her thickest armor was 4.5 inches, and she carried 508 men.

Akagi was a Japanese-built gunboat, four years old (sadly no reliable representation of her is at hand). Speaking of her and her three sisters, Watts and Gordon state: "All were completed with schooner rig, but this was later reduced and just before the war with China", the *Akagi* was fitted with a fighting top to the foremast and a crow's nest to the mainmast. This vessel could always be distinguished from her sisters because she was fitted with a forecastle. The 8.2-inch old-pattern Krupp gun was forward, with the 5.9-inch aft. She was unarmored and carried 104 men.

Kotaka was an eight-year old British-built torpedo boat. Two of the torpedo tubes were fixed to fire ahead, the other four paired amidships and aft on trainable mounts. Her machinery was protected by one-inch armor, her complement is not a matter of record. *TB1* through *TB4* were British-built, fourteen years old. Two torpedo tubes were fixed in the bow and one trainable amidships. They were unarmored. *TB5* through *TB14* and *TB16* through *TB19* were three to four years old, French-built; one tube in the bow, one trainable amidships; unarmored. *TB15* and *TB20* were French-built, two years old; tubes on a trainable twin mount aft; unarmored. *TB21* was French-built, three years old; fixed tube in the bow, trainable amidships and aft; unarmored. *TB22* and *TB23* were German-built, two years old; one fixed tube in the bow, two trainable amidships and aft; unarmored.

NAVAL BATTLES OF THE SINO –JAPANESE WAR, 1894 - 1895

THE BATTLE OF ASAN, JULY 25, 1894

(This engagement is occasionally still known as the Battle of Phung-Tao,)

Japanese provocations in Korea having mounted alarmingly, the Chinese determined to dispatch a formation of elite troops to that area. The British steamer *Kowshing* was engaged as a troop transport, and in due course set out loaded with 1100 men, accompanied by *Chi Yuan* and *Huang Yi*. The style of the escort was distinctly loose, but after all this was peacetime.

Proceeding along the west coast of Korea on the misty morning of July 25, the cruiser and the gunboat were well ahead of *Kowshing* and out of visual contact with her. Approaching from the opposite direction, the Chinese made out the Japanese First Flying Squadron under the command of Rear-Admiral Tsuboi. This squadron was composed of four cruisers, *Yoshino*, *Naniwa*, *Taka Chi* and *Akitsushima*. Whatever unease this may have occasioned aboard the Chinese vessels, it was not sufficient to prompt them to come to action stations.

Thus they were at considerable disadvantage when, disregarding formalities *Naniwa* opened fire upon them, and the remainder of the squadron followed suit. *Kuang Yi* got her QF guns into action and made a run for it, but *Chi Yuan* had the worst of it from the start, and quickly struck her colors without having opened fire.

While the remainder of the First Flying Squadron moved off in pursuit of *Huang Yi*, *Yoshino* closed in to accept the surrender of *Chi Yuan*. When *Yoshino* had approached to within two hundred yards, with the rest of the Japanese disappearing into the mist, the Chinaman, (apparently finding these odds more to his taste) suddenly opened fire and discharged a torpedo.

Japanese accounts maintain that this attack "disabled" *Yoshino*, but I find no evidence that she was seriously damaged. However this may have been, *Chi Yuan* succeeded in her escape, making away in the

91

direction she had come. She passed *Kowshing* but made no attempt to alert the steamer to the Japanese attack.

Huang Yi fought bravely enough, but she was no match for the three larger warships. At length she was pounded to wreckage by shellfire, *Naniwa* and *Akitsushima* taking the main part in her destruction, and was run ashore to avoid sinking, a total loss. Eighteen men survived out of her crew of 120. The mist and the confusion of battle had by now caused the Japanese to lose touch with one another, and in fact the First Flying Squadron did not reform during the remainder of the day.

Kowshing came up about an hour after the battle, when *Naniwa* intercepted her and commanded her to halt and surrender. A party was put aboard to advise her to follow the Japanese cruiser into port, an order to which her European officers agreed readily enough. As soon as the boarding party had cleared, however, the Chinese troops asserted their unwillingness to comply with the Japanese instructions. As they physically impeded attempts to get *Kowshing* under way, the Europeans signaled the Japanese, informing them of the situation aboard. As the troops prevented the Europeans from leaving the ship, *Naniwa* opened fire after four hours of fruitless negotiation.

Five 10.5-inch shells struck *Kowshing* at point-blank range, and she sank stern first after thirty minutes. Her captain and first officer, with one other white seaman, were picked up from the water by the Japanese. A German major attached to the Chinese army swam the two miles to shore, accompanied by 160 of the Chinese troops. Two of *Kowshing*'s three lifeboats had been launched successfully, but they were shot to pieces by Japanese machine-guns. Of the remaining 940 Chinese troops, all were slain, either going down with the ship, drowning, or being machine-gunned in the water.

Later in the afternoon, *Tsao Chiang* was sighted proceeding along the coast. *Akitsushima* overhauled her and demanded her surrender, which she forthwith tendered without resistance. (*Tsao Chiang*, renamed *Soko* and rearmed with two 12-pounder (76-mm) QF guns, was commissioned in the Imperial Japanese Navy, in which she served until 1903.

It was not until seven days later, after the Chinese Army had been engaged and defeated at Songwhan by the forces of General Oshima, that on August 1, 1894 a tardy declaration of war was issued.

THE BATTLE OF THE YALU RIVER

SEPTEMBER 17, 1894

(Eastlake and Yoshiaki refer to this battle as: "the naval engagement of the Yellow Sea, better known by the style of the 'Fight off Haiyang'," but in English-language texts it has long since been known exclusively as the Battle of the Yalu River.)

For a time the Chinese navy was virtually restricted to port, in governmental reaction to the defeat at Asan. At length, however, the rapid advance of the Japanese Army in Korea having made necessary the reinforcement of the Chinese by sea. A group of five transports carrying 5000 soldiers was therefore dispatched under convoy of twelve warships and two torpedo boats, the entire force commanded by Admiral Ting Juchang. Arriving without incident at the mouth of the Yalu River on September 16, the main body of the fleet stood watch while *Ping Yuen*, *Kuang Ping* and the torpedo boats accompanied the transports upriver. The entire contingent of troops was landed without loss during the afternoon and the following morning.

September 17, dawned clear and fine, the sea glassy under a light breeze. The Chinese crews were exercising at anchor when a cloud of heavy black smoke was sighted bearing SW. Sources differ as to whether the sighting took place at 0950 or 1000. (The Japanese sighting of black smoke bearing NE by E is given as 1050; this suggests that, as was later their custom, their clocks remained set to Tokyo time and were thus an hour ahead of the Chinese local time. Times given henceforth are Japanese.)

Knowing by their coal smoke that the enemy fleet was approaching, Ting at once gave General Quarters; the Chinese had steam up and were under way by 1100 (their time). They adopted line abreast in the following order (port to starboard) *Kuang Chia*, *Chi Yuan*, *Chin Yuan*, *King Yuan* , *Chen Yuen*, *Ting Yuen*, *Lai Yuen*, *Ching Yuan*, *Yang Wei*, *Chao Yung*. It will be remarked that Ting had concentrated his four most powerful vessels in the center of his line, with the less battle worthy units at either extremity. *Ping Yuen*, *Kuang Ping* and the torpedo boats trailed along some distance astern, at the anomalous little armored cruiser's best speed of around seven knots. (Note that *Kuang Ping*, with the Chinese squadron's only medium-caliber QF guns, is thus absent at the initial encounter. No clearer illustration could be desired of the Chinese lack of faith in, or ignorance of, the new ordnance.)

The Japanese fleet was organized into two semi-independent squadrons as was their practice here and in future wars as well. The First Flying Squadron, commanded by Rear Admiral Tsuboi with his flag in *Yoshino*, comprised the flagship, *Takachiho*, *Akitsushima* and *Naniwa*. These were as the name implies, faster warships, protected cruisers. The Principal Squadron, with the commander-in-chief, Admiral Ito in direct command in *Matsushima*, comprised the flagship, *Chiyoda*, *Itsukushima* and *Hashidate*. The Japanese approached the engagement in line ahead in the order given above. *Hiei* and *Fuso* ("obsolescent ironclads," as Potter and Nimitz rightly describe them) were assigned station at the rear of the line, but they were able to keep up only with difficulty. *Akagi* and the armed merchantman *Saikyo Maru* ("transformed into a cruiser for the time being," Eastlake and Yoshiaki inform us disarmingly) were of negligible fighting value, and trailed on the disengaged side of the formation.

The two fleets came into visual contact at 1140. The Japanese were in single column, the Flying Squadron in the lead. The Chinese were still in line abreast, but by this time the formation was ragged, with trailing flanks, in fact a shallow, somewhat irregular wedge. Potter and Nimitz suggest that Ting may have intended to execute Ships Right on contact to form line ahead toward Port Arthur and safety. If this is so, the slackness of Chinese station-keeping thwarted him, for the formation was too loose for such a maneuver to be effectively performed. At any rate, no such order was given.

It is entirely realistic to believe, however, that Ting may have intended from the beginning to fight in line abreast. Certainly the configuration of his ships would have encouraged this tactic. The two battleships could fire all four of their 12-inch/20 guns straight ahead only, the theoretical cross-ship fire from the off-side turret for a four-gun broadside for which the ships were designed being in practice unfeasible because of excessive blast damage. The armored cruisers *King Yuan* and *Lai Yuan*, the only other truly battle worthy Chinese units, mounted the two 8.2-inch/55 Krupp guns in barbette forward, with the two 5.9-inch in sponsons amidships, thus also being able to maintain fire with all four guns only on a forward bearing. The other ships also displayed a greater or lesser degree of forward-firing bias.

At 1205 Admiral Ito made "Close the enemy," and the Flying Squadron began at once to draw ahead of the slower Principal Squadron. To a lesser extent and less rapidly, *Hiei*, *Fuso*, *Akagi* and *Saikyo Maru* , none of them truly fit to stand in the line anyway, began to slip out of station and straggle astern because of their inferior performance.

The Japanese approached from the port bow of the Chinese line, traveling at twice their enemy's speed. Clearly, Ito anticipated Ships Right or Left by Ting, which would have brought the two fleets into parallel lines ahead. The Flying Squadron bore straight in toward the center of the Chinese line, ready to form line in either direction in conformity with the enemy. The anticipated maneuver failed to occur, the Chinese continuing to bear down in line abreast. Observing this, the Flying Squadron, with the range down to 6000 yards, veered to port and made for the Chinese starboard wing.

At this moment, about 1250, Admiral Ting made "Commence firing" and the flagship *Ting Yuen* opened the engagement with one of her 12-inch guns; (The concussion of firing knocked Ting and every man on the bridge to the deck; the admiral himself had to be assisted below. This speaks of very lax gunnery drill indeed; plainly the officers on the bridge had never before been aboard when the main armament was fired.) The remainder of the fleet joined in at once, subjecting the still silent Japanese to a slow, deliberate, and largely inaccurate bombardment.

For five minutes, Tsuboi held his fire as his ships bore down diagonally across the front of the advancing Chinese. As the range came down to 3000 yards, optimum range for their 6- and 4.7-inch quick-firers, the Japanese also opened a brisk fire, concentrating on *Yang Wei* and *Chao Yung*, the nearest vessels of the enemy. The Flying Squadron briefly doubled the Chinese starboard wing; then, sighting the smoke of *Ping Yuen* and company away to northward, the Squadron bore up to confront this new, unidentified menace, temporarily losing contact with the main Chinese force.

The Principal Squadron followed in the wake of the Flying Squadron, opening fire in turn, but the Chinese were closer now, and the least battle-worthy Japanese units, the four obsolescent stragglers, were in the greatest danger. Visibility was rapidly deteriorating as well, as the clouds of coal and gun smoke hung drifting listlessly in the near-calm. The big guns of the Principal Squadron concentrated upon the Chinese battleships, while the quick-firers sought to strip away the weakly-armored flanks of the enemy formation. At 1258 a 12.6-inch shell from *Matsushima* clipped the mainmast of *Ting Yuen*, greatly impeding the flagship's signaling ability, and the Chinese formation rapidly ceased to function as a coordinated whole.

The major vessels of the Principal Squadron doubled the starboard flank of the enemy in sufficient time, and proceeded to steam across the rear of the Chinese line. The stragglers were in genuine peril as the Chinese bore in upon them, however. Only the slowness of Chinese fire, the defective Chinese ammunition, and the concealment afforded by the standing smoke preserved them from destruction at this time. *Saikyo Maru* made off to the north, signaling the Flying Squadron that the Japanese rear was being annihilated. Tsuboi at once came about, abandoning *Ping Yuen*, and steamed back to the rescue. In the meantime the Principal Squadron had crossed the Chinese rear and taken up position on the port quarter of the enemy formation. The ironclads had extricated themselves from their predicament: *Akagi* passed around the starboard wing of the Chinese formation, and *Fuso* did likewise, although the latter sustained shell damage. *Hiei* was obliged to run between *Ting Yuen* and *Chen Yuen*, which is to say directly through the middle of the enemy formation, in order to regain her own fleet. This the elderly corvette accomplished without damage, firing ineffective broadsides from her outmoded guns into the two battleships in passing.

While in the narrow sense the recall of the Flying Squadron had thus been unnecessary, in fact Tsuboi's return was most fortunate. The danger posed by the ill-conceived *Ping Yuen* and her consorts was negligible; indeed they never caught up and took no part whatever in the battle, and the Flying Squadron now fell in opposite the Principal Squadron, on the Chinese starboard quarter. The battle had now assumed its ultimate configuration, and caught between two fires, the ragged Chinese line quickly began to disintegrate.

There is considerable confusion as to the precise sequence and detail of the action beyond this point; it has even been said that no coherent account of the later stages of the battle exists. While this may be strictly true if we retain in mind the basic situation the two Japanese squadrons hanging like wolves on the flanks of the dwindling Chinese wedge, into which to fit the various specific incidents, we may readily piece together a comprehensible and perfectly adequate account of the battle's progress.

Despite the one-sidedness of the ensuing fight, it ought to be noted that the Chinese were not behindhand in gallantry. Their weapons were obsolete their leadership incompetent, their training and discipline inadequate, and their ammunition seriously defective, but they stood to their guns and banged away doggedly enough for all that.

Yang Wei, already damaged earlier in the battle, soon was hopelessly afire and stood out of action. Eventually the lightly-built ship ran ashore on Talu Island to avoid sinking. (Emerson and Miller alone maintain that, in fleeing the contest, *Chi Yuan* rammed and sank the burning *Yang Wei*. They seem motivated in this curious opinion by their hatred for *Chi Yuan*'s commanding officer, "the wretched Fong." They deplore his "cowardly stratagem at Asan, where it will be recalled that he struck his colors but subsequently assailed *Yoshino* and escaped." While strictly speaking this was contrary to naval usage, it must be remembered that *Chi Yuan* had just been attacked by foreign warships in time of peace, a circumstance which renders high-toned castigation of the victim's subsequent conduct hypocritical. As no other source confirms the ramming of *Yang Wei*, although the questionable McElwee vaguely mentions a collision, and as all other sources which mention her, place *Yang Wei* firmly aground on Talu Island rather than accidentally sunk, we may assume that Emerson and Miller slander Captain Fong.)

Ching Yuan and *Lai Yuan* were also soon burning, but managed to keep station after a fashion, at least retaining visual contact with *Chao Yung*. *Yang Wei*'s sister and the only other ship of that ill-starred class, was not so fortunate. Holed below the waterline, she fell behind. *Chao Yung* fought to the end, her 10-inch guns booming out from their curious square gun houses. One of the shells slammed in aft on *Matsushima*, striking the barbette of the 12.6-inch gun, but it was a dud and did no vital damage. Overwhelmed by shellfire, dead in the water and blazing, *Chao Yung* shortly slid down by the stern and sank.

On the other flank, *Chi Yuan* quickly fled, as had been mentioned, and *Kuang Chia* joined her in precipitous retirement toward Port Arthur. (*Kuang Chia* seems, however, despite her brief participation, to have sustained considerable damage, for Conway reports her "beached near Port Arthur, becoming a total loss.") *Chih Yuan* and *King Yuan* , although damaged, managed to maintain station for the moment.

Now, and indeed throughout the remainder of the action, the Chinese battleships consistently attempted, by tight, maximum-speed turns, to bring their bows (and thus their full firepower) to bear on the vessels of the Principal Squadron (albeit without success, for the Japanese vessels were speedier and handier). The results of this maneuvering were twofold: damaged Chinese vessels were quickly left at a distance from the squadron; and the four cruisers of the Flying Squadron were often left at a distance from the foe, being obliged to sweep fairly widely in echelon to regain contact, for visibility continued to deteriorate in the drifting banks of smoke. It was during such a sweep that the Flying Squadron made contact with *King Yuan* . Already much damaged and fallen hopelessly out of station, the German-built armored cruiser stood in firing gamely, but she was the only enemy vessel in sight, and the four Japanese cruisers concentrated their fire upon her. Unequal to the weight of shellfire which now enveloped her, the blazing *King Yuan* exploded with great violence, rolled over to port and sank. In much the same fashion *Chih Yuan* was caught up. The crews of her thinly-shielded guns swept away by the storm of quick-firer projectiles that burst over her, the British-built cruiser nevertheless valiantly bore in, attempting to ram *Yoshino*. Of course she had no chance; as the range closed the quick-firers ate away at her superstructure and even the big guns scored, tearing great holes into her hull. *Chih Yuan* listed to starboard and went down in the vortex of shellfire at 1530, never having seriously menaced her erstwhile target.

Meanwhile the Principal Squadron had been constantly engaged with the sadly diminished Chinese formation (for only *Lai Yuan* and *Ching Yuan*, fighting their fires and serving their guns as best they might, had succeeded in keeping station on the battleships). The Japanese, holding the range at 2500 yards, had concentrated their fire mainly upon the two capital ships. The superstructures of both battleships were a shambles, but their fourteen-inch-thick compound armor had proved impervious to the Japanese ordnance, and their vitals were unharmed.

At 1420 a 12-inch shell struck *Saikyo Maru* . The shell failed to explode, like so many of the Chinese projectiles, but it cut all rudder controls, and the concussion of its impact shook the merchantman to the keel, and severely strained her relatively fragile frame. She had been of no real value anyway, and now she was out of the fight once for all, although by superior seamanship she was brought into port.

In the course of evolutions, the Principal Squadron permitted itself to come up abeam of the Chinese, and seeing an opportunity, *Ting Yuen* suddenly veered to port and ran down upon *Matsushima*, seeking to ram. The Japanese opened the range, and every gun concentrated upon the approaching battleship. Wreathed in smoke, which trailed out behind her, her ravaged superstructure sparkling under the rain of quick-firers, her 12-inch Krupp guns booming steadily and deliberately, the squat German-built *Ting Yuen* came on, relentless and indestructible. A shell from one of the great 12.6-inch Canet rifles crashed aboard, spraying a great burst of flame and debris, but even this failed to slow or deflect the Chinaman's onset.

Yet it was futile in the end. With their superior speed, the Japanese kept the range open, and at length the battered *Ting Yugn* swung around and returned toward her squadron. At this moment (it was 1550, and off to the north *Chih Yuan* was just sinking), the Chinese flagship fired a parting round from one of her 12-inch guns which scored on *Matsushima*. The dud shell penetrated the starboard quarter, detonating the ready ammunition of the 4.7-inch secondary battery, passed through the sickbay, and finally stove in and cracked the twelve-inch-thick barbette, coming to rest in this position and inextricably jamming the main gun. The starboard secondary battery was burned out, 100 men were dead or wounded out of a crew of 360,

100

and the ship was afire. Ito transferred his flag to *Hashidate*. Southworth, who has throughout given a curiously jumbled and disjointed account of this engagement, finally tips his hand at this point with the following extraordinary and utterly erroneous account:

> Among the wounded was the captain of the *Chen Yuen*. Philo N McGiffin succeeded to the command. As an undergraduate at Annapolis, McGiffin had once engineered the prank of simultaneously discharging six ornamental cannons on the Academy grounds, by remote control. Seeing some virtue in this practical joke, he had installed a buzzer system on this Chinese battleship to permit the four 12-inch guns to be fired together. His gun crews had received training in this primitive salvo fire. Now he ordered "Cease fire" and had the steersman follow an erratic course, as though the *Chen Yuen* were partly disabled. As he hoped, an enemy cruiser closed for the kill-- Ito's flagship *Matsushima*. As she drew closer, the four 12-inch guns followed her, training in unison. Choosing a moment when all four had an unobstructed bearing, McGiffin pushed the button and all four big guns fired together. At least two of the shells hit the *Matsushima* squarely. Armor and hull were pierced, the forward turret put out of action, a 4.7-inch battery silenced, and more than fifty casualties caused. Badly hurt, the *Matsushima* went limping out of action. Admiral Ito hastily transferred his flag to the sister-cruiser *Hashidate*. (Southworth)

The fact that must now be introduced, although I would otherwise have preferred to omit it, is that the unfortunate Captain McGiffin's behavior had initially prevented him from serving in the United States Navy and he entered the less demanding Chinese service. He was given 297 demerits while a midshipman at Annapolis, 3 shy of the number which would have prevented his graduation. He returned in due course to his native land, and in 1905, he ended his own life. He has few apologists and those that he does possess admit that his insubordinate behavior at Annapolis was a significant reason for his failure in the US Navy (Davis in Selected References). Clearly what we have here is a sample of his grandiose account of the Battle of the

Yalu River, the only noteworthy event in which he ever participated, and he has contrived this story to maximize his own role therein. He contradicts all other sources, so plainly his story should not be allowed to prevail. At any rate, I would point out that simultaneous fire by the guns of the main battery is of no particular value unless the ship possesses a central fire control ranging positioner, which *Chen Yuen* did not, and which even Captain McGiffin did not claim to have installed. In short, if you cannot guarantee one hit, you cannot guarantee four.

(I would, as I said, have preferred not to enter into the painful and unfortunate history of Captain McGiffin, but as his unsupported accounts have come to pretend to the dignity of history, I was left little choice.)

By his own account, Ito had by now determined that the remaining Chinese ships were indestructible with his weapons, and thus he drew off, maintaining a distant visual contact with the enemy, and firing fell off and ended gradually. (I dare say that Admiral Ito was also a bit shaken by the admittedly unnerving devastation of his flagship by a single defective round from the 12-inch guns of the enemy battleships, and this may have played its part in his decision to allow the fighting to lapse at this point. Also, during the flag transfer the Japanese had doubtless fallen into some confusion allowing the Chinese, whose ammunition was virtually exhausted, to open the range. Upon establishing himself in *Hashidate*, Ito was probably more inclined to allow the existing situation to persist than he would have been to initiate the break himself.)

The Japanese tailed the enemy at long range until, at 1750 with darkness falling, Ito broke off contact entirely, fearing an assault by Chinese torpedo boats. None too soon for the Chinese, for *Chen Yuen* was down to three 12-inch shells (although her casualties were merely 14 killed and 25 wounded, despite the estimated 400 shell hits she sustained) and *Ting Yuen* was nearly as depleted. The battered Chinese ships, left in meaningless possession of the field, straggled together. *Ting Yuen*, *Chen Yuen*, *Lai Yuen* (still fighting her stubborn fires) and *Ching Yuan*, retiring in the direction of Port Arthur, caught up and carried along the laggard *Ping Yuen* and her company as well.

Watts and Gordon, *Weapons and Warfare* and (surprisingly) Conway unite in asserting that, in Conway's words, "after taking part in the Battle of the Yalu, *Akitsushima* assisted the cruiser *Naniwa* to sink the Chinese cruiser *Kuang Chi*"(sic). In fact no such episode occurred, and the Chinese cruiser *Kuang Chi* never existed. What seems to have happened is that *Kuang Chia* and *Kuang Yi* have become confused with one another. Both, after all, were run ashore to avoid sinking, becoming total losses, and their names are much alike. Thus the circumstances of the destruction of *Kuang Yi* on July 25 appear to have been erroneously transposed to the loss of *Kuang Chia* on September 17.

On the morning of September 18, the Japanese mistakenly swept toward Weiheiwei in search of the Chinese remnants, so the retirement was effected without incident. The Chinese fetched up at Port Arthur, where they remained for several weeks, until a landing by the Japanese Second Army on October 24, rendered the place untenable and they fled to Weiheiwei. "Unaccountably," as Dupuy and Dupuy remark, the Japanese failed to hinder their passage of the Pohai Straits.

THE ASSAULT UPON WEIHEIWEI

JANUARY 17 - FEBRUARY 12, 1895

With the arrival of the fugitives, the Beiyang Fleet was once more concentrated at Weiheiwei; after the surrender of Port Arthur to the Japanese (19 November, 1894) that was also the only port still open to them. In the harbor at Weiheiwei were *Ting Yuen, Chen Yuen, Lai Yuen, Ping Yuen, Ching Yuen, Chi Yuen, Kuang Ping, Fu Lung* and thirteen other torpedo boats. The entire Japanese fleet was committed to the blockade of Weiheiwei, and as soon as the army could make preparations, an amphibious assault against the port was mounted, for with the neutralization of Weiheiwei and the Chinese ships, Japanese dominance of Korea and the Yellow Sea would be complete.

On January 17, 1895 the First Flying Squadron covered the landing of a diversionary force twenty miles west of Weiheiwei. On Jan 19 the real assault went in twenty-four miles east of the harbor at Jungcheng. The unopposed landing of the Japanese Second Army was completed in five days. Ashore were 52,000 men, 6000 horses, and supplies and munitions sufficient for six weeks' operations under the command of General Oyama.

The blockade had meanwhile been the occasion of desultory, long-range exchanges of fire between the Japanese warships and the two Chinese battleships inside. Damage was minor, although casualties were incurred on both sides. Moving off on January 24, well ahead of any possible intervention by the Shantung Army, Oyama's force crossed frozen rivers in bitter cold and by January 29, were in position to assault the eastern batteries of Weiheiwei. The fleet bombarded the forts, then lifted their fire as the attack went in. The infantry swept into the batteries and held them, seized the station for the electric mines blocking the harbor, and removed the boom closing the eastern entrance. The fighting continued next day, and by January 31, the entire land shore of Weiheiwei Harbor was in Japanese hands, although Chinese troops remained on the islands. Chinese losses to this point are estimated at 2000 men.

There now began a ponderous ballet that was to continue during daylight hours for the next twelve days. The captured shore batteries, supplemented by army heavy artillery, took the Chinese vessels under fire; but the guns were manned by soldiers, untrained in the art of ranging on warships. Despite their relatively confined situation, the Chinese ships dodged this fire with remarkable success, avoiding any serious damage whatever for several days.

Admiral Ito, perhaps irritated by the prolonged and hopeless resistance of the all-but-captive Chinese, resolved to attack the ships at night with torpedo boats. The first operation took place on the night of February 4-5, in cold so intense that men froze to death at exposed stations. The night was dark. During the entry into the harbor. *TB22* ran hard aground on a reef and became a total loss, although her crew was rescued. *TB8* went aground also, and although she was not destroyed, she was nevertheless eliminated from the night's operation. The same fate befell *TB14*, wrecked with eight men dead in the frigid sea.

The remainder of the boats entered the harbor and engaged the Chinese warships. It was a wild melee in the icy dark: like many night actions, it was so confused as not to be susceptible to detailed reconstruction. Although the Japanese pressed home the attack with the utmost vigor, only one torpedo hit was secured, *TB25* succeeding in her assault upon *Ting Yuen*. The battleship was mortally hurt, but due to her watertight compartmentation (a novelty at the time of her construction) she kept afloat until morning and her crew got away with little loss. (The seeming paucity of results is not surprising when we reflect that the torpedo is an intricate and sensitive mechanism. Even during World War II there were numerous instances of torpedo malfunction attributable to extreme cold; how much more detrimental it must have been to the relatively fragile and far crankier torpedo of 1895.)

The next night the boats went in again. Perhaps it was a little warmer or the crews may have benefitted from the previous night's practice; at any rate success was greater. *Lai Yuan* was torpedoed by *Kotaka* and capsized, her upper works sticking in the mud of the harbor floor. Trapped crewmen could be heard "knocking and shrieking for days before they expired."

TB25 scored again, sinking the disarmed training ship *Wei Yuan*. (This ship was originally a sister of *Tang Ching*.) A torpedo hit was secured upon *Ching Yuan* as well, immobilizing her. On the other hand, *TB19* was badly mauled by fire from the Chinese warships.

Admiral Ting must have sensed that the end was near. His crews were wearing down under the strain of the two sleepless nights; depletion of his fuel and ammunition must have been of immediate concern, for how could he have replenished? At any rate, he ordered his torpedo boats on the evening of February 7, to "break out to the open sea;" a quixotic mission. Two of them "reached safety"; exactly where, or the details of their escape, I have been unable to discover. Eight others were driven ashore and destroyed, and *Fu Lung* and three other boats taken by *Yoshino* and the ships of the First Flying Squadron. (The captured boats were later incorporated into the Imperial Japanese Navy. *Fu Lung* was renamed *Fukuriu*, and the other three became Third Class Torpedo Boats *TB26*, *TB27* and *TB28*. *TB28* was stricken from the Fleet List in 1902, but the rest served on through the Russo-Japanese War.)

On February 9, to compound Ting's already hopeless dilemma, the shore artillery finally began to get the range. The disabled *Ching Yuan* was sent to the bottom by a 10-inch shell from a captured shore battery, and *Chen Yuan* was sunk in shallow water by the fire of army siege artillery. Ting arranged the capitulation of Weiheiwei, to take effect on February 12; he took particular care to secure guarantees of the personal safety of his crews. This done, he retired to his quarters, "swallowed a lethal dose" of opium and ended his life.

In China, Admiral Ting was held responsible for the annulation of the Chinese fleet. In Japan, however, he apparently gained considerable respect for choosing suicide rather than surrender.

The disposition of the surrendered warships was as follows:

Chen Yuan was refloated, refitted, rearmed with two 6-inch QF in place of the 5.9-inch, an additional two 6-pounder and eight 5-pounder QF; renamed *Chin Yen.* in the Imperial Japanese Navy.

Ting Yuen was refitted with British guns, one 10.2-inch, two 6-inch QF, eight 3-pounder QF, and renamed *Hei Yen.* .

Chi Yuan was renamed *Sai Yen* and incorporated into the Imperial Japanese Navy. In "the late 1890s" she was rearmed with eight 3-pounder QF in place of the original 3-inch.

Kuang Ping was rerated a Third Class Cruiser, renamed *Ko Hei*, and incorporated into the Imperial Japanese Navy.

Ko Hei was removed from the Fleet List in 1897, but the other three served on into the Russo-Japanese War, where we shall encounter them once more.

___

In addition to the ships that Japan gained, were also territorial acquisitions, principle among them being Port Arthur. This strategic harbor was named for William C. Arthur, a Royal Navy Lieutenant, who first mapped it in 1860 during the Opium Wars. It was seized by Japan following the Sino-Japanese War, but Russia "coerced" a treaty with China and occupied the port two years later. It would become the focal point of the Russo-Japanese War a decade later.

The Opium Wars (First and Second) were fought between 1839 and 1860, and pitted the European nations of Britain, France and Russia, and the United States against the Quin Dynasty of China. The issue was Chinese sovereignty and foreign rights to trade opium, among other goods, in China. The immediate precipitant was the Quin Dynasty's attempt to outlaw the opium trade in China when they became alarmed by the growing number of opium addicts. China was overwhelmed in the wars and the result was the opening of China to western trade, the legalization of opium in China, and several leases to European nations most notably Hong Kong to the British.

There were battles involved in this struggle, some involving warships, but none were of such a scale as to warrant discussion here and of course took place prior to the period under discussion. None-the-less the reader may wish to bear this episode in mind when reading about the US acquisition of Guantanamo Bay in Cuba (page 131).

The drug trade and the eventual control of large areas of China has obvious parallels to the geopolitical situation today.

CONCLUSIONS REGARDING THE SINO-JAPANESE WAR

It will be apparent that, as the first fleet action on the open sea since the Battle of Lissa twenty-eight years before, the Battle of the Yalu River commanded intense international interest. Many of the "lessons" learned were of the elementary sort which one thinks common sense ought to have taught anyway: for example, that it was unwise to employ wood for the interior fittings of warships, as had been the practice, since explosive shells would set fire to it.

A great deal of perplexity was occasioned among tactical theorists, for the Yalu seemed directly to contradict the verdicts of Lissa. The line ahead had now defeated the line abreast, and the gun had established its superiority over the ram. This was the more striking since maps of the opening phases of the two battles are virtually identical, yet that disposition which triumphed at Lissa was roundly defeated at the Yalu. Having said this much, it will be as well fully to compare the two battles.

It is instructive to begin our analysis with those features which the two engagements have in common, rather than with their more obvious differences. I think it may be said that in both cases a resolute commander with a definite and coherent plan of battle imposed his will upon a materially superior foe who through negligence or adverse circumstance offered battle haphazardly. The disparate tactics of the two victors are the result of the particular situations. Tegetthoff quite properly mistrusted the ability of his smooth bore ordnance to disable the Italian ironclads; therefore his desire was to bring on a close-range melee as rapidly as possible, so that the enemy might be rammed, and at any rate the advantage of his superior rifled cannon would be minimized. For this reason he adopted the most compact possible disposition, "balling his ships together," so to speak, and hurling them at the enemy.

Ito, on the other hand, had the utmost confidence in his artillery, and therefore wished to keep the range open and pound his enemy to pieces without subjecting any of his own units to undue risk. Thus he conformed to the traditional line ahead, as the formation permitting the most effective control of the entire squadron by the

108

commanding officer. He himself assumed the commodore's customary station at the head of the central division, the position offering the commanding officer optimum visibility. (The practice of operating the divisions of the fleet semi-independently had such happy results in this instance largely because Ito was blessed with a capable and like-minded subordinate in Vice-Admiral Tsuboi, who in the heat of battle comprehended his commander's purpose and fulfilled it without need of communication. This sort of rapport is not common, and we shall see that often enough such an arrangement of divided command tends rather toward rancor and inefficiency.)

In extenuation of Admiral Ting it has been remarked that a number of foreign "advisors" held positions subordinate to him; the reluctance with which these white nineteenth-century naval officers would execute the orders of a Chinaman (especially if they doubted the sense of them) may be readily conjectured. This in part explains the fact that while individual ships fought well, attempts at coordinated maneuver were dismal failures, although the extremely variable performance of the various units of the squadron must not be underestimated in this connection. The early disruption of *Ting Yuen*'s signal halyards will also be recalled. (It has upon occasion been extrapolated from the circumstance of Ting's being thrown to the deck by the initial shot of the battle and subsequently assisted below, that he was thenceforth incapacitated and "the fleet left leaderless." Even if it were accurate, this rather melodramatic interpretation is of limited validity: as insufficient leadership had been demonstrated to weld the Beiyang Fleet into a unified fighting force prior to the first shot, then even had he been at the top of his form Admiral Ting could do little enough once hostilities had actually commenced.) Indeed, Ting is a melancholy figure. Despite his mediocre abilities, he was true in the end to his duty. His defense of Weiheiwei has quite rightly been described as "the sole speck of honor which China garnered" in the course of the war. There is little room for such ambivalence of feeling with regard to Admiral Persano. The sole redeeming aspect of his performance is his initial reluctance to lead his fleet into action, insofar as this betrays lack of confidence in his own abilities.

As regards the ram, it may be seen with the advantage of hindsight that the spectacular sinking of *Re d'Italia* which so captured the imagination of contemporary thinkers was in fact a very special case. The loss that was truly significant for the future was the explosion of *Palestro*.

Concerning the controversy between big guns and medium-caliber QF ordnance, the decision was moot. Quick-firers had performed well against lightly-protected warships, as everyone had conceded they would, but failed to disable or even to seriously inconvenience the Chinese battleships. On the other hand, the big guns had fared no better against the capital ships; the one or two hits which, at their current rate of fire, were all they could reliably be expected to score, were clearly insufficient to "smash" a modern battleship. This was a major surprise. Artillerymen as they so often do when issued improved equipment during a period of extended peace, had greatly overestimated the effectiveness of their weapons, It had been believed that naval engagements would be very short, since no ship would be able long to withstand the devastating power of modern explosives. In fact, as we have seen, the event established that armor had more than kept pace with ordnance, and that a battleship was still all but impervious to shellfire.

Lieutenant W. S. Sims USN was intelligence officer aboard the cruiser *Charleston* on the China station at the time, and I quote Admiral Morison's summation of his report:

> His ultimate conclusions were that the Chinese had had very fine equipment and ordnance which they failed to use at all intelligently. Battleships had proved themselves superior to cruisers, but in this particular case the Chinese had been so incapable of maneuvering properly that they had been defeated by inferior forces.

Except to remark once more upon the crippling deficiencies of Chinese ammunition, I think we may allow these conclusions to stand.

NAVAL BATTLES OF THE
SPANISH-AMERICAN WAR, 1898

GENERAL INTRODUCTION

Rebellion in the Spanish colony of Cuba had long concerned the government of the United States. In December, 1897 the U S consul at Havana, Fitzhugh Lee, alarmed at increasing mob violence in the city, requested that a warship be placed on alert to proceed thither should the need arise, "to furnish tangible and moral support to his office." *Maine* was accordingly detailed for this duty, and placed on two-hour alert at Key West.

Maine, completed in 1895, was rated a second-class battleship; she had originally been classified as an armored cruiser, and it was thus that European naval annuals continued to designate her. She displaced 6682 tons, was 319 feet over all, with a speed of 17 knots. She was armed with four 10-inch/50, six 6-inch/30, seven 6-pounder QF, and eight 1-pounder QF guns, and four 14-inch torpedo tubes. Her thickest armor was 12 inches, and she carried 374 men.

Consul General Lee summoned his back-up on January 12, 1898; but the situation stabilized and he cancelled his request before the ship could sail. The government was nevertheless unsettled, and desired that *Maine* should be situated so as more readily to intervene in the event of crisis. After some debate it was decided that *Maine* would pay "a friendly visit" to Cuba, reviving a custom of reciprocal US-Spanish naval visits which had been suspended for three years. The Spanish government gamely announced that the cruiser *Vizcaya* would therefore visit New York; their genuine emotions in the face of the transparent ruse may be readily imagined.

Maine was accordingly ordered out on January 24, and arrived in Havana on the morning of January 25. At the conclusion of her visit, Consul General Lee requested that her recall be countermanded, and that she remain indefinitely in Cuban waters. His request was granted, but at 2140 February 15, 1898 *Maine*'s forward magazine detonated and she settled rapidly to the harbor bottom with the loss of 252 of her company.

It is virtually beyond dispute that *Maine* was destroyed by an underwater explosive charge deliberately positioned underwater against her hull. US public opinion, not unnaturally, blamed the Spanish authorities, and war was thenceforth inevitable. On April 25, Congress issued a declaration of war, predating the document to April 21, in order to legalize the blockade of Cuba that had been in effect from the earlier date.

As the war from this point divides itself into two completely distinct theaters, Philippine and Caribbean, we shall treat the two in sequence, The Spanish-American is the most exhaustively documented of any war between 1866 and 1913; histories of the war are readily available. Therefore we shall be able once again to concentrate on what happened during the battles. Those desiring to learn of Commodore Dewey's vigorous and far-seeing preparations which were crowned by victory at Manila Bay, or of the altogether less felicitous pursuit of (or search for) Cervera's squadron, which was conducted by Admirals Sampson and Schley, (although it was worthy of the Keystone Kops), will have no difficulty in discovering these details elsewhere. Neither do I have any intention of discussing the disgraceful and acrimonious dispute that smoldered on for years afterwards between Admiral Sampson and Admiral Schley, and by which these two mediocre officers demeaned themselves and besmirched their service.

INTRODUCTION TO THE SHIPS AT MANILA BAY

SPANISH VESSELS

Castilla was a Spanish-built vessel, dating from 1882. She had been designed originally as an ironclad and was thus "heavily and clumsily built" despite having been completed in wood. Given her material, the ram bow was a preposterous affectation. In 1885 she was fitted with four 5.9-inch guns mounted in sponsons and two 4.7-inch BL. Her guns were of Krupp manufacture, and her complement was 392 men.

Reina Cristina was completed in 1888 and armed with six domestically produced Hontoria 6.4-inch guns mounted in sponsons, and her 6 and 3-pounder quick-firers manufactured by Hotchkiss. She was "unarmored" but "built on the cellular system, with twelve water tight compartments." She carried 570 men.

Velasco, Don Juan de Austria and *Don Antonio de Ulloa* were virtual sisters, despite the variant armament of *Velasco* which was equipped with four 6-inch and two 3-inch guns. The two sisters were armed with four 4.7-inch and four 6 pounder guns. *Velasco* completed in 1882 the others, not until 1889. *Velasco* was British-built, whereas *Austria* and *Ulloa* were made in Spain to the British designs. They carried 173 men each.

Isla de Cuba and *Isla de Luzon* were British-built, completed in 1887. They were steel-hulled, with a maximum armor thickness of 2.5 inches. Armament was six 4.7-inch and eight 6-pounder quick-firing guns. Their complement was 164 men.

General Lezo was an "iron-hulled boat with light schooner rig and a single funnel", completed in Spain in 1885. She carried 98 men. (Her sister, *El Cano*, is sometimes said to have been present at Manila Bay as well.)

Margues del Duero, "iron-hulled schooner-rigged gunboat with one funnel and very prominent ram bow", completed in France in 1875. Her crew numbered 98.

The armament of these ships at the time of the battle is not at all clear since many of the guns were either not functioning or had been moved ashore.

113

UNITED STATES VESSELS

Olympia, completed in 1895, was designed as a fast commerce raider. She was the first US warship to conform to the classic pre-dreadnought outline, the 8-inch guns in twin turrets fore and aft and the 5-inch in broadside battery. Her thickest armor was 4.5 inches, and her complement as a flagship was 447 men. She can be seen as of this writing at the Independence Seaport Museum in Philadelphia.

Baltimore was constructed to a British design, the work of Sir William White, and the ship completed in the US in 1890 (ironically the design had been previously submitted to and rejected by the Spanish government). The 8-inch guns were sponsored fore and aft, with the 6-inch in broadside. The thickest armor was 4 inches, and the crew numbered 586.

Boston was designed and launched by the notorious John Roach, and was completed at New York Navy Yard after Roach's trial for corruption and shoddy workmanship. The forward 8-inch gun was offset to port, with the after gun to starboard; the 6-inch were sponsored in broadside. The thickest armor was 2 inches, and the ship's company numbered 284. She completed in 1887.

Raleigh was a second-class cruiser completed in 1894. The single 6-inch gun was forward; two of the 5-inch were mounted abreast on the main deck aft, with the rest in broadside sponsons. Her thickest armor was 4 inches, and she carried 312 men.

Concord was completed in 1891. The 6-inch guns were in pairs, slightly sponsoned fore and aft, and in full broadside sponsons amidships. She was originally classed as a cruiser. Her conning tower was protected by 2 inch plate, although she was otherwise unarmored; her crew numbered 187.

Petrel was a rather curious~looking little boat, completed in 1889. The four 6-inch were mounted in broadside sponsons; she was unarmored and carried 158 men.

114

A brief word about John Roach as an example of an entrepreneur in the growing shipbuilding industry seems appropriate here. He emigrated from Ireland and founded Roach and Sons, a major ship building company operating from 1864 until 1885 when it went into receivership. At its peak, it was the largest shipbuilder in the United States and the largest employer outside of the railroad industry. In many ways, the story is characteristic of armament sales and acquisition in governments at that time and even to this day.

The building of warships, in this period in particular, was rapidly changing, difficult to understand, and difficult to anticipate. This was true for the designers and builders and even more so for the congressional committees of the United States who made decisions regarding the purchasing of these weapons. Couple this with the fact, as emphasized in this writing, that the success or failure of a design had ultimately to wait upon the next battle. There was obviously ample room for error. Unfortunately, there was also room for manipulation and graft. The administrations prior to President Cleveland's election in 1885. were among the worst in this regard, *Grantism*, named for President Grant (1868 – 1876), had become a synonym for graft and corruption, this in-spite-of the fact that Ulysses Grant was himself probably among the most honest of presidents.

Grover Cleveland was elected in part as a backlash to this situation. He was nicknamed "Grover the Good", in recognition of his honesty and abhorrence of corruption among public officials. His campaign slogan was "Public Office is a Public Trust". He was also the only Democrat elected to the office of the president between the election of Abraham Lincoln and William Howard Taft (1860 to 1913). Cleveland accused Roach and Sons of shoddy workmanship and graft and canceled the navy contracts that supported the company driving it into receivership in 1885.

The truth of the accusation is hard to determine. Roach was a heavy contributor to the Republican Party and benefited greatly from naval contracts awarded to his company by Republican administrations. His competitors were among his most vocal critics. The shoddy workmanship was not well substantiated, and some argue it was merely innovative. For example, *Columbia*, a passenger steamer, was the first ship utilizing a dynamo and the first structure, other than

Edison's Menlo Park laboratory in New Jersey, to use incandescent light bulbs.

The corruption chargers were probably justified. His ship designs were criticized by the *The Engineer*, a British engineering journal, but as we shall see, many of the design problems were the result of requirements imposed by the government, such as the persistent use of cylindrical "pill box" turrets and small coal bunker capacity. Graft was a different issue.

A congressional hearing established that the Secretary of the Navy, George M. Robeson, had received bribes, although the investigation was not able to establish any charge against Roach. One of his major journalistic attackers was Charles Dana, editor of the *New York Sun* and his persistence was sufficient to end the government's contracts with Roach.

The business did survive but Mr. Roach did not. He died of a "mouth cancer" in 1887. His son, also named John Roach, continued the business until 1908. In all they built 179 ships, including 10 warships for the United States Navy and one for Spain.

Interestingly, Grover Cleveland developed mouth cancer in 1893 as well, but survived after surgical removal of the tumor, done secretly on a friend's yacht, the *Oneida*. Only one reporter, E. J. Edwards of the *Philadelphia Press*, found out about the surgery, even verifying with one of the surgeons, but Cleveland denied the story, fearing that it would worsen the depression the country was experiencing, and it was ultimately Edwards who was disgraced. The truth came out decades later when Dr. William Williams Keene, one of the surgeons, disclosed the story. Times have changed, and apparently so has the definition of honesty. Today it might be Mr. Edwards who would become the hero and Cleveland who was disgraced.

One more diversion before returning to the battles, Ulysses Grant also died of a head and neck cancer, attributed to his alcohol and tobacco use. So it is that all three of the principles in this saga were stricken by the same disease. Strange perhaps, but merely a coincidence, unless they all were liberal users of alcohol and tobacco. While lacking documentation, it is none-the–less possible in this age when the use of both by gentlemen, was expected.

THE BATTLE OF MANILA BAY MAY 1, 1898

Subig Bay having been entered and found empty of Spanish warships, Commodore Dewey was virtually certain that his enemy would be found in Manila Bay, and set his course thither. At 2330 April 30, the entrance to the bay had been reached, and the US squadron passed into the channel; the ships were in the order *Olympia* (flag), *Baltimore*, *Petrel*, *Raleigh*, *Concord*, *Boston*, the revenue cutter *Hugh McCulloch*, and the recently purchased colliers *Nanshan* and *Zafiro*.

The warships made their way safely through the relatively narrow passage between Corregidor and El Fraile, but as *McCulloch* passed near the two islands soot in her funnel caught fire and she belched a cloud of flame into the night. A rocket arose from Corregidor in a flaring red parabola; before it sputtered out into the sea it had been answered by another from El Fraile. Momentarily, at 0015 May 1, 1898, the battery on El Fraile fired three rounds toward the US ships. No damage was done, and fire was instantly returned; *Boston* fired one 8-inch shell, *Concord* two rounds of 6-inch, *Raleigh* two 5-inch, and *McCulloch* three or four 6-pounder shells. The Spanish fire ceased forthwith; Corregidor had not fired at all, although a sentry there later reported that the ships were plainly visible. The Americans, doubtless a little astonished at their good fortune, proceeded in silence through the darkness. It was Dewey's intention to arrive off the city of Manila at daybreak, for he firmly expected to find the Spanish ships moored under the city's forts; therefore he restricted his speed to six knots.

At first light Dewey detached *McCulloch*, *Nanshan* and *Zafiro*, and stood in toward Manila with his warships in single line ahead. The morning was already quite warm, misty and flat calm. The enemy was not where Dewey anticipated to find him, but fairly rapidly the mist began to burn off and the Spaniards were detected, drawn up in the inlet at Cavite.

Admiral Montojo had determined to fight his ships at anchor, not without reason. *Castilla* could not get under way; her boilers were bad, furthermore she leaked uncontrollably if her engines were started. Most of *Uloa*'s machinery was ashore under repair, and *Velasco* was in worse shape, with neither her engines nor her guns aboard.

The ships had formed up at Cavite, rather than in the more favorable position directly beneath the batteries of Manila, because the humane but utterly pessimistic Montojo desired to spare the city from the destruction he was already convinced was about to engulf his fleet. As a result, his ships were too distant from the majority of the shore artillery to be effectively supported by it. *Velasco* was moored closer inshore, and *Lezo* was in Bakor Bay with the armed merchant cruiser *Mindanao* and possibly *El Cano* as well. The remainder of the ships stood in an irregular crescent line in the order *Cristina* (flag), *Cuba*, *Luzon*, *Ulloa*, *Austria*, *Castilla* and *Duero*, their port broadsides facing the enemy. All the ships in the line save *Ulloa* and *Castilla* had steam up and springs on their cables; the immobile *Castilla* was protected by lighters filled with sand and stones which were secured to her port side.

It was 0500 as the Americans discerned the location of the Spanish line, and the ships at once steered for the enemy, *Olympia*, *Baltimore*, *Petrel*, *Raleigh*, *Concord*, and *Boston*, 400-yard intervals between ships. At 0506 an electrically-detonated mine went off about 1000 yards ahead of *Olympia*, throwing an awesome column of water high into the air. Dewey did not deflect his course, and no further mines detonated.

At 0515 the batteries of Cavite and Sangley Point opened an inaccurate fire at long range, and the Spanish ships followed suit. The Americans continued to close, their guns silent. The shells of the defenders fell for the most part ahead of and beyond the approaching ships, which sustained virtually no damage. It was not until 0541, with the range down to 4000 yards, that *Olympia* made "Fire when convenient" and opened with one of her forward 8-inch guns. At the same time she turned to starboard to pass roughly parallel to the Spanish line; her 5-inch opened, and the following ships did likewise as soon as they made the turn and their guns bore.

As the Americans passed along the Spanish line at a stately six knots, a launch flying a large Spanish flag was seen to approach *Olympia* as if to attack with torpedoes. A veritable hurricane of 6-pounder shells burst all about, but the launch, seemingly impervious, made a leisurely pass well within torpedo range of the American flagship, turned away, running along calmly in the eye of the storm of artillery, and eventually went aground beneath Sangley Point. In fact she was no torpedo boat, but a privately owned steam launch carrying Filipino servants to market; with "not bravery but incredible rashness" she had attempted to follow her daily routine, battle or no. Her passengers survived, although one shell did pierce her steam cylinder. (So strong was the impression that she had been a torpedo boat that not only did she continue to draw unauthorized 6-pounder fire throughout the remainder of the action. She was also included in Dewey's original list of destroyed Spanish warships. Her name was given as "*El Correo*", which is merely to say in Spanish "*The Ferry*", and two 4.7-inch guns were erroneously, and quite fantastically, attributed to her. In his revised list, *El Correo* was omitted, presumably after her true nature was ascertained, but the marginally less questionable *El Cano* took her place.)

The big guns concentrated mainly on *Cristina*, with her conspicuous admiral's flag, and *Castilla*, whose high white-painted side stood out like a billboard. The Spanish gunners in turn tended to concentrate on *Olympia*, both because of Dewey's flag and because as the leader she was less obscured by smoke than the ships in her wake.

Reaching the end of the Spanish line, *Olympia* came right about, the others following in her wake, and passed back down, a little closer to the enemy and thus inboard of the standing smoke from the previous pass. The Spanish anchorage was by now largely obscured by smoke; some of the Spaniards had slipped their cables and were maneuvering aimlessly, masking the fire of their comrades. The Spanish fire was hot and vigorous, albeit still almost entirely inaccurate. The Americans' fresh starboard batteries were in action now, while the Spaniards' port batteries were for the most part still engaged.

By 0620 the Americans had returned to the starting point of their promenade, and *Olympia* came about once more to pass for a third time along the Spanish line, still closer to the enemy. This was the most nerve-wracking firing pass for the Americans, The enemy was entirely shrouded in smoke, and the gunners were firing "by memory", with no opportunity to verify their marksmanship. They did not seem to be doing well, for the Spanish fire had not by any means slackened. On the other hand it was obvious that the Spaniards were getting the range, perhaps because the Americans tended to be silhouetted against the smoke bank from their previous runs. *Olympia* received three hits, one on the foremast, and two near the bridge, one of the latter passing through the grating upon which the commodore stood. *Raleigh* was obliged to cut away her shot-up boats, lest their debris foul her screws. *Boston*'s rigging was set alight; although brief and superficial, the blaze was quite spectacular, and the other ships entertained the most exaggerated idea of *Boston*'s injury. Further, the breech-plugs of *Boston*'s 8-inch guns had expanded with the heat of continuous fire and could only be closed with difficulty. *Baltimore*'s guns and ammunition hoists had begun to evidence a variety of malfunctions, which seriously reduced her rate of fire. Near the end of the Spanish line, her gunners struggling with their recalcitrant paraphernalia, *Baltimore* shuddered under a medium-caliber shell hit; eight men were slightly wounded and damage was slight, but the glut of flame and great bursting cloud of brown smoke was seen throughout the squadron.

Undaunted, the flagship led her charges around for another firing pass, yet closer to the enemy. It was during this run that the battle began to go visibly in favor of the Americans. A shot severed *Castilla*'s forward cables, and the current slowly swung her bow toward the enemy, continuing to bear her around until her unshielded starboard side faced the Americans. She was already badly afire, and only one of her guns was operational. Her captain gave "Abandon ship" and her crew went over the side and swam ashore, leaving 15 men dead on board. As *Boston*, at the end of Dewey's line, began to pass down the Spanish formation, *Margues del Duero* "pointed her long ram toward Sangley Point" and attempted to escape by passing astern of the Americans. She was taken for a torpedo boat, and a brisk fire concentrated upon her, driving her back among her consorts with her engines badly damaged and her pathetically obsolete muzzle-loading smooth-bore cannon dismounted. *Don Juan de Austria* made a brief sortie also, but the concentration of fire she experienced as soon as she advanced out of the smoke was more than she could bear, and she retreated with 15 men dead.

It was 0700 by the time *Olympia* swung around for the fifth run-past, and Dewey at this time veered rather more toward the enemy, bringing the range down briefly to 1500 yards. Seeing the enemy this close, and realizing that with his ship already afire it was now or never,

Admiral Montojo made "Ram the enemy" and *Reine Cristina* steamed up at top speed upon *Olympia*. The same storm of shellfire that had greeted *Duero* and *Austria* now burst upon the flagship, for his subordinates had ignored Montojo's signal, and *Cristina* attacked alone. She was already burning forward; a shell struck the bridge and destroyed the helm; another destroyed the sickbay; an 8-inch projectile penetrated the magazine, which by good fortune was flooded in time; another of 6-inch caliber shattered her stern. The hand-steering station was thus rendered untenable, and the battered cruiser continued out of control. An 8-inch shell crashed through her stern and raked her main deck, producing a great gust of flame through every port and skylight. *Reina Cristina* at length ran aground in the shallows behind Cavite, and giving "Abandon Ship", Montojo secured his flag and transferred to *Isla de Cuba*. Losses were given at the time as 52 dead and 150 wounded, but when the ship was raised in 1911 the remains of 80 men were found in the sick bay alone. By this time *Don Antonio de Ulloa* had sustained a hit at the waterline, and three of *Isle de Luzon*'s guns were inoperable.

Imagine the astonishment and delight of the Spaniards then to see at 0755 *Olympia* turn precipitously away and lead her squadron out into the bay out of range. Commodore Dewey, underestimating the damage he had inflicted upon the enemy, largely because they were obscured by smoke and the continuing vigor of their fire was deceptive, had received a thoroughly alarming report to the effect that only fifteen rounds of 5-inch ammunition remained aboard *Olympia*. He ordered the report verified, but by the time the correction reached him, that fifteen rounds per 5-inch gun had been fired, he was already standing out into the bay and firing had ceased. He resolved to confer with his captains, while allowing the men to breakfast at their stations. Apprehension turned to incredulity, and then to elation, as ship after ship reported negligible damage, no dead, plenty of ammunition; the only casualties were *Baltimore*'s eight wounded. Further, as the smoke drifted clear of the Cavite anchorage, it could be seen that the enemy was in sorry state. *Castilla* was utterly engulfed in flames, her masts falling in swirling tornadoes of sparks as they watched. *Reina Cristina* stood thoroughly afire some distance from the anchorage. Such of the rest as could get underway were moving into the shelter of Bakor Bay behind Cavite.

By 1100 the men were fed, and *Baltimore* stood in toward Sangley, considerably in advance of her consorts. At 1116 she resumed action, firing one of her forward 8-inch guns into the abandoned, blazing *Reina Cristina*, which promptly exploded with great violence. *Baltimore* stood on until she was within 1000 yards of the Sangley fortifications, elevating her guns to fire up into the embrasures. The cannonade was tremendously destructive, the shells penetrating up under the steel glacis plates and flipping them back onto the guns. The white flag arose over Sangley Point.

Meantime *Olympia* led *Concord*, *Boston*, *Raleigh* and *Petrel* in that order back toward Cavite anchorage, to find the battle all but won. The ships in Bakor Bay had complied willingly with Montojo's final "Scuttle and abandon your ships."; the crews then following their admiral ashore. *Don Antonio de Ulloa*, immobilized by the absence of her engines, was the only ship which still stood to meet the enemy. She proceeded to prove that it was not her disposition to retreat in any case, opening a brisk fire upon the overwhelmingly superior forces of the Americans, which she maintained until she was shot to pieces, capsizing and going under about 1130. The guns of Cavite Arsenal continued to fire, until a fortuitous 8-inch shell demolished the dining room of the commanding officer's home. The battle having thus become personal, that individual forthwith surrendered the arsenal and all the works.

By 1230 the ships had withdrawn once more into the middle of Manila Bay, save the shallow-draft *Petrel*, which was ordered into the shallows of Bakor Bay to destroy the remaining-Spanish vessels. She sank *Duero* by gunfire, although the gunboat was abandoned, presumably electing to take no chances with her vicious-looking ram in the constricted bay. Boats put out to *Don Juan de Austria* and *Isla de Cuba*, the only other Spanish warships still afloat. Discovering the two to be abandoned, the Americans scuttled them as well. *Petrel* also took two large tugs, three launches and a number of small boats, which she towed back to the squadron in an impromptu triumphal procession.

Total US casualties for the battle were *Baltimore*'s eight wounded. *Montojo* originally gave his losses as 101 dead and 280 wounded, but as 66 of the wounded were beyond help and died within forty-eight hours, the figures are customarily given as 167 dead and 214 wounded. (The 80 skeletons in *Reina Cristina*'s sickbay would seem to give the lie to these figures anyway, but that is as it may be.) The U S squadron scored 242 hits out of 5859 shells fired, an accuracy rate of slightly less than 2.5 percent.

Isla de Luzon, *Isla de Cuba* and *Don Juan de Austria* were refloated, extensively reconstructed, and incorporated into the US Navy. *Luzon* "served with the Louisiana and Illinois Naval Militia 1905-1918". *Cuba* was sold to Venezuela in 1912, renamed *Mariscal Sucre*, and served until 1940. *Austria* "served with the Michigan Naval Militia 1907-1917"

INTRODUCTION TO THE SHIPS AT
SANTIAGO de CUBA

JULY 5, 1898

I will state at the outset that it is my intention from henceforth to refer to true turrets, hooded barbettes, and fully-shielded mounts indifferently as turrets. The technical distinctions are of no importance in the present work, and the use of the term, turret, to designate any fully-enclosed rotating gun-mount, although strictly erroneous, has become conventional.

SPANISH VESSELS

The three *Vizcaya*s were the entirety of their class. They were built in Spain, *Teresa* completed in 1890, *Vizcaya* and *Oguendo* in 1891. They were the result of Spain's desire for a "force of much larger cruisers intended for mobile defense and to protect the sea lanes" connecting her with her few remaining colonies. They were stately and impressive ships, with a maximum armor thickness of 12 inches. They had a high unprotected freeboard, however and the 5.5-inch guns were mounted in broadside on the open upper deck. Turret armor was only 1.5 inches thick. "In order to avoid excess heat building up in the 'tween- deck spaces, the upper decks were made entirely of wood, without the customary steel deck beneath." This was to prove their fatal flaw. They carried 484 men.

Cristobal Colon was Italian-built, a cruiser of the *Giuseppe Garibaldi* class. Conway calls them "a very successful class," but *Colon*'s efficiency was diminished by the fact that at the time of purchase, the Spanish had such urgent need that it was not possible to mount her main armament of two 10-inch/40 guns. She commissioned on May 16, 1898 and sailed almost at once to join Cervera's squadron. Her steel armor was 4.8 inches thick and she carried 543 men.

Furor and the slightly larger *Pluton* were of almost identical appearance. These "torpedo-boat destroyers" (as they were initially known), despite their poor performance at Santiago, were a type of the future nevertheless. The concept was of British origin (and in fact these specific ships were of British manufacture). The desire was for a ship fast and agile enough to catch torpedo boats, and with sufficient gun power to destroy them. Having achieved this, it was an inescapable corollary that the destroyers should carry torpedo tubes as well, since their small size and greater speed and seaworthiness fitted them to deliver torpedo attacks against the enemy battle fleet. Indeed, they were marginally better suited to this duty than were the torpedo boats they rendered obsolete. *Furor* completed in 1896, *Pluton* in 1897; they were unarmored and carried 67 men.

UNITED STATES SHIPS

Iowa was a modification of the *Indiana* class; she completed in 1897. Although the 12-inch main guns seem to Conway a retrogression from the 13-inch of the *Indiana*, in fact the larger gun was cumbersome and unwieldy, and the slightly smaller caliber of the guns on *Iowa* was more than compensated by the greater manageability and increased accuracy. The 8-inch turrets were further from the main turrets than those of the *Indiana*, thus minimizing blast interaction between the main and secondary mounts. The 4-inch tertiary guns were quick-firers, replacing the less numerous old-style 6-inch aboard the earlier ships. The additional displacement made it possible to add a raised forecastle deck that extended back to the rear edge of the superstructure, thus improving her seakeeping qualities over those of her predecessors. *Iowa*'s thickest armor was 17 inches, and her complement was 654.

The *Indiana*'s were the first modern battleships constructed for the US Navy, *Indiana* completing in 1895 and *Massachusetts* and *Oregon* in 1896. Most of their shortcomings can be traced to the restrictions placed upon them by the reactionary and "coast-defense"-minded Congress. They were unfit for extended sea service, due to their low freeboard and small bunker capacity.

> The Navy Department preferred the old cylindrical 'pill box' turret, and in order to secure the maximum elevation . . . the designers had to place the 13-inch mounting close to the edge of the turret . . . As a result, the ships listed sharply when the guns were trained on the beam . . . It was found that the 8-inch guns could not be fired within 10° of the centerline as the blast from their muzzles concussed the men in the sighting hoods of the 13-inch turrets. (Fitzsimons, *Weapons and Warfare*)

While strikingly unusual in appearance, these were still quite handsome ships, and the bristling tiers of turrets gave them a formidable aspect. Their thickest armor was 18 inches, and they carried 656 men.

Texas was a second class battleship built to a British design and completed in 1895. She was not a successful ship, being too small and weak to support her 12-inch guns; these could only fire at a restricted angle on their own broadside, since end-on or cross-ship fire resulted in prohibitive blast damage. She was a stubby and unattractive vessel, bearing no close resemblance to any other warship, although the shielded 6-inch guns fore and aft under the overhang of the superstructure gave her a vaguely Russian look. Her thickest armor was 12 inches, and she carried 508 men.

New York, the first US attempt at an armored cruiser, completed in 1895 and was an impressive and magnificent-looking ship. (*Maine* had been classified briefly as an armored cruiser, but this was in no genuine sense accurate.) Four of the 8-inch were in twin turrets fore and aft, with the remaining two in single sponsons on either broadside amidships. Her thickest armor was 10-inch nickel steel, and she carried a crew of 566.

Brooklyn, completed in 1896, was according to Conway "a ship of unusual appearance", and in fact she was much like a Frenchman. The 8-inch were in four twin turrets, one fore and one aft and one on each broadside amidships. The thickest armor was 8.5 inches, the complement 581.

New Orleans would join the United States fleet. She was a protected cruiser that had been laid down in England as *Amazonas* to a Brazilian order, but as war loomed she was purchased on the slips by the United States. *New Orleans* was armed with six 6-inch/50, four 4.7-inch/50, ten 6-pounder and eight 1-pounder guns, and three 18-inch torpedo tubes. Her thickest armor was 5.5 inches, and she carried 366 men. She was, as we shall note later, using smokeless propellant powder, a novelty at this time.

THE SIEGE AND BATTLE OF SANTIAGO de CUBA
MAY 28 - JULY 3, 1898

Admiral Cervera entered the harbor of Santiago de Cuba on May 19 1893 in *lnfanta Maria Teresa*, with *Cristobal Colon, Almirante Oquendo*, *Vizcaya*, *Pluton* and *Furor* in company. Already present in the port was the cruiser *Reina Mercedes*. She was a sister to Admiral Montojo's flagship at Manila Bay, *Reina Cristina*; she was in poor condition, and most of her artillery was removed and emplaced ashore to supplement the harbor's forts, four of whose guns dated from 1663. (The third ship of the class, *Alfonso XII*, was in Havana where she had been damaged in the explosion of the *Maine*; she took no part in the war.)

On May 26, with his ships as well reconditioned and replenished as the meager facilities would permit, Cervera posted *Cristobal Colon* in the harbor mouth as a sentry while he pondered his next move. His despondent reflections were overtaken by events. At 1750 May 28, the dilatory Admiral Schley, at length took station off Santiago, where his superiors had been trying for almost a week to send him. His Flying Squadron consisted of *Brooklyn* (flag), *Iowa*, *Massachusetts*, and *Texas*. Later in the day the Flying Squadron was joined by the recently purchased *New Orleans*. It was not until the morning of May 29, three days after *Colon* had been stationed, that she was observed anchored in the entrance channel.

On May 31, with *Colon* still in plain sight, Admiral Schley at long last decided that he ought to at least open fire upon her. He formed *Massachusetts*, *Iowa* and *New Orleans* up in line ahead and steamed past the anchored Spaniard, firing at a range of 7500 yards. *Colon* replied with her 6-inch guns; no damage was done to either side. The gunnery practice of *New Orleans*, with her fine new Armstrong guns and smokeless propellant powder, was nevertheless most impressive to the Americans, whose domestic ordnance was markedly inferior. The following morning, *Colon* upped anchor and withdrew out of sight into the harbor. A few hours later, Admiral Sampson, Admiral Schley's immediate superior, arrived to take command, his flag in *New York* with *Oregon* in company.

On June 2, Admiral Sampson assigned blockading stations to his ships. The two armored cruisers and four battleships were positioned in a semi-circle from four to six miles from the entrance, with smaller craft (mostly requisitioned yachts armed with small quick-firing guns) patrolling closer inshore. (The US Navy was utterly deficient in small craft; thus numerous private vessels were taken into service, crewed for the most part by volunteers and reservists.)

Sampson also quickly concluded that the narrow entrance channel to the cul-de-sac harbor of Santiago would readily lend itself to the sinking of a block ship to "cork the bottle" in which Cervera's squadron was confined. The collier *Merrimac* was selected as being both large and expendable enough; she was fitted with ten external electrically-fired explosive charges, evenly spaced down both sides of her hull. A crew of seven volunteers under the command of Lieutenant Richmond Hobson was chosen to take her in. The plan was that, at the narrowest point of the channel, she should port her helm and run her bow onto the shore, at the same time firing her charges, so that the current would swing her stern over against the opposite bank as she sank. Thus she would come to rest perpendicularly across the channel, denying egress to the enemy within.

It was originally intended that she go in on the night of June 2, but various delays prevented this, and she was obliged to wait for darkness to fall on the night of June 5. It was a cloudy night, with lightning flashes visible behind the hills of Cuba. *Merrimac* ran in without incident to within a quarter-mile of her intended resting place. At this range, however, she was sighted by the defenders and a heavy and accurate fire was opened upon her. The shore batteries and *Vizcaya*'s 5.5-inch guns engaged her, and she was also subjected to the massed musketry of two regiments of soldiers, who were lined up along both sides of the channel. She could not put her bow onto the bank because her steering gear was destroyed by shellfire; the attempt was made to sink her in the proper place, but only two of her scuttling charges detonated. The wiring of the other eight had been shot away, so that she filled too slowly and drifted well beyond the narrows before sinking at a point where she was no particular obstruction to passage. All seven of her crew survived, clinging to the catamaran they had provided for their escape. They were unable to clear the wreckage, since continued rifle fire made it impossible to get aboard the catamaran, and they were obliged simply to hang on to it to keep afloat, with only their heads above water.

At daylight, a launch commanded by Admiral Cervera himself (who insisted upon coming personally as a token of his admiration for their courage) took the seven prisoner. They were well (I had almost said lavishly) treated, and a boat was sent under flag of truce to inform Admiral Sampson that the seven had survived and were in Spanish hands. The only casualties of the episode were four Spanish soldiers slain in error by the rifle fire of their comrades during the cross-fire in the channel.

At about this time a suspicious vessel sighted upon the high seas, although it proved not to be one of Cervera's cruisers, prompted the realization that no one on the American side knew for certain that the entire Spanish squadron was blockaded in Santiago. Accordingly, arrangements were made with insurgents ashore. An officer of the fleet was landed, climbed a hill and counted the ships within the harbor. The cruisers were all there, which was a relief; but Admiral Sampson was also apprised for the first time of the presence of the two destroyers. Sampson had already been uneasy about the possibility of a Spanish escapade under cover of darkness; now the threat of night torpedo attack sharpened his anxiety. Accordingly he resolved upon an audacious and unorthodox tactic: with the fall of night, the ships would close to within 2000 yards of the entrance and train their searchlights upon it. Picket boats from the ships also patrolled right up to the harbor mouth during the hours of darkness. "A damned impertinence", a British officer who saw it characterized the procedure, and one could scarcely wish a more emphatic illustration of the self-confident panache with which the US Navy operated throughout. (Night illumination of the harbor mouth was adopted on June 8, and remained in force through the night of July 4. It is occasionally said that Cervera might well have escaped had he sortied at night, but those who propose this have forgotten the searchlight drill, or were never aware of it.)

Meanwhile, on June 6 the shore batteries had been taken under fire by the blockaders. The ships formed up and moved in at 0640 under lowering skies, with rain squalls blowing continually across the surface of the sea. Firing commenced at 0730 and ended at 1000.

Spanish losses were 19 men killed and 56 wounded; one gun was totally destroyed. The warships suffered no losses or damage, and henceforth regarded the forts with contempt. The batteries were, however, quite sufficient to prevent minesweeping in the channel proper, and until the mines could be cleared, the ships could not enter the harbor to force action upon Cervera. Admiral Sampson accordingly, on June 7, advised the War Department that 10,000 troops might in forty-eight hours capture the batteries; he requested that a suitable force be dispatched to do so, in order that the operation might be swiftly concluded. After a slow start, the ships of the blockading force were rapidly shaking down into a unified and coordinated fighting force. The same could not always be said for the swiftly commissioned yachts:

On June 9, one of these yachts reported sighting a squadron of Spanish ships that was believed to include a battleship. This eventually proved to be a group of five US auxiliary vessels who thought the yacht was a Spanish torpedo boat and fired at her. The yacht in turn was too far away to hear the gun discharges and thought the muzzle flashes were signal lights. While this episode highlights the deficiencies of these quickly recruited additions to the US Navy, some performed remarkably well, notably the *Gloucester* as we shall see.

Since it was becoming obvious that an extended naval presence off Santiago might be required, it was a matter of some urgency that a base ashore be acquired for coaling and maintenance. Accordingly, on June 10 a force of Marines comprising 652 officers and men with four 5-inch field guns was put ashore at Guantanamo Bay, a natural harbor suited for development into a fleet facility. The initial landing was unopposed, but the first Spanish counterattack developed that same afternoon, and fighting continued for five days. At length, the Marines secured the only fresh-water wells in the area; deprived of these, the Spanish ceased to operate against the harbor, and coaling from colliers commenced immediately. Guantanamo Bay was eventually leased to the United States in perpetuity by Cuba in 1903 and reaffirmed again in 1934 to "enable the United States to maintain independence of Cuba".

Among the ships supporting the landing was the *Texas*, commanded by Captain John Philips. Philips maneuvered Texas two and one half miles up the bay to engage Spanish fortifications making success possible. We shall see him again in the climactic battle of the campaign, exercising remarkable competence there as well. He was an intensely religious man, objecting to the bombardment of Spanish forts on a Sunday and succeeded in having the bombardment postponed until Monday. Another noteworthy incident occurred when the *Massachusetts* accidentally discharged a shot that arced over *Texas.* Philips responded with the humorous signal to Captain Higginson on *Massachusetts*: "Good line shot, but a little high."

On June 15th the blockade was reinforced by the arrival of the experimental dynamite-gun cruiser *Vesuvius*. She was a ship of 929 tons, 252 feet 4 inches long over all, with a speed of 20 knots. Her armament consisted of three smooth-bore 15-inch Zalinsky pneumatic guns. These were 55-foot long tubes from which a 250-pound dynamite bomb was projected by compressed air. The bombs were very similar to aerial bombs of our own day, even to the point of being finned to ensure striking at the correct angle. The tubes were fixed three abreast in the extreme bow at 18° elevation and pointed along the line of bearing. They were aimed by aiming the ship; range was altered by varying the air pressure. Accuracy was very poor, and the maximum range was 3200 yards. Her conning tower was protected by 1-inch plate, and her complement was 70. As it was judged unwise to commit *Vesuvius* within range of the enemy's shore batteries in daylight (one hit among her dynamite bombs and she would have gone up like an ammunition transport), she was employed only after dark. She was active every night until the conclusion of the blockade, closing to within 1500 yards of the entrance and silently lobbing her projectiles along the beams of the searchlights. Although she did virtually no material damage, the totally unheralded and extremely violent detonation of her bombs created great alarm among the Spanish gunners. Her performance was adequate, but *Vesuvius* was nevertheless withdrawn from service by the end of the year, and no further dynamite-gun warships were contemplated. This was not so much on account of any specific failing, but rather because apart from her novelty she provided nothing more than an inferior duplication of capabilities already possessed by the conventional ships of the fleet.

On the morning of June 14 *New Orleans* closed to within 2000 yards of the western fortifications and silenced them in twenty minutes of firing, sustaining no damage herself. This action was undertaken because news had been received that General Shafter, commanding 16,000 troops in 32 transports had sailed to support the fleet. It being thus established that the forts had not been reinforced and still presented no difficulties to warships offshore, the arrival of the Army was awaited with confidence. The flotilla arrived on June 20, escorted by *Indiana*; the battleship joined the blockade although her usefulness was limited by the advanced dilapidation of her boilers.

It had been Admiral Sampson's intention when he requested troops that a bombardment by the fleet to silence the works should be followed up by immediate landings below the works, which could, he reckoned, be carried by storm with minimal losses. Mine clearance could then take place, after which the ships could enter Santiago Harbor and destroy Cervera's squadron at anchor.

General Shafter, who was not subordinate to Admiral Sampson, had other ideas. The landing places chosen by the Navy seemed to him dangerous; furthermore, he was well aware that his might be the Army's only substantial campaign of the war (as indeed it proved to be), and he had no intention of conducting it in the shadow of the Navy. They had garnered all the glory thus far, but now it was to be the soldiers' turn.

At Shafter's insistence, the troops were put ashore at Daiquiri, an unsuitable beach eighteen miles east of Santiago. The landing, supported by units of the fleet, was completed by June 24 with the loss of two men drowned. The Spanish made no attempt to intervene. Rather than march along the coast to attack the batteries, Shafter directed his forces inland to attack the city of Santiago proper. After a brief skirmish at Las Guasimas, the troops attacked the Spanish outworks at El Caney and San Juan Hill on July 1, carrying them after a day of bitter fighting. It was in this engagement that Teddy Roosevelt and his Rough Riders saw their famous action and leaped instantly into the limelight and fame, largely the result of Teddy's audacious personality. The fact that the "Buffalo Soldiers" (African-American troops) of the 10th Cavalry and 24th Infantry Regiments had actually done much of the heaviest fighting did not diminish Teddy's accomplishment as recorded in the US press.

133

General Shafter, meanwhile, was confined to his bed by illness miles from the fighting, and was influenced by unduly pessimistic reports which described the men as utterly spent and the position as dangerously vulnerable to counterattack. He seems at this point to have completely lost his nerve, and he spent July 2 simultaneously informing the War Department of his intention to withdraw five miles and dig in, while at the same time demanding that Admiral Sampson force the harbor of Santiago with the fleet in order to create a diversion and "save the Army", and advising the commander of the Spanish forces that if the city were not surrendered at once, he would reduce it with his artillery.

Admiral Sampson's incredulity may be imagined, more especially as Shafter's message was particularly insolent ("I cannot understand why the Navy should not face gunfire and risk losses as well as the Army."). It was by now late in the day, so Admiral Sampson arranged to come down the coast in the morning of July 3 in order to confer with General Shafter and hopefully to make some sense of his army counterpart's seemingly irrational maneuvers and demands.

From the other side, the situation wore a startlingly different aspect. The Spanish commander, his outer works irreparably breeched and the majority of the defenders dead at their posts, felt that the end was near and was inclined to treat the surrender demand with the greatest seriousness. The Governor-General in Havana had been urging a sortie upon the reluctant Cervera for some time, "to maintain the honor of Spain". In any case, he was ordered, if the fall of the city seemed imminent, the fleet must at all costs attempt to escape; total destruction was to be preferred to ignominious surrender.

Admiral Cervera had contested these orders as vigorously as he could, writing to his government in Madrid, "The absolute and certain result will be the ruin of all the ships and the death of the greater part of their crews," and characterizing the proposed sortie as a "horrible and useless hecatomb". His representations had been in vain, and now, with the surrender of Santiago virtually inevitable, the time had come for him to do his duty and obey the hateful command.

His plan was that *Maria Teresa* should ram *Brooklyn* (the only ship of the US squadron that ought to have been able to match speed with the Spaniards while the rest of his ships broke west along the coast. Thus, if they were badly damaged, they could run ashore, and hopefully some of their men would be saved.

July 5th dawned fair and warm, with a light northeasterly breeze. *Massachusetts* was detached at dawn to coal at Guantanamo. Admiral Sampson made ready to confer with General Shafter, and *New York*, making "Disregard movements of flagship", left station at 0855 and proceeded east down the coast to carry him to the rendezvous. This left on the blockade (from west to east) *Brooklyn*, *Texas*, *Iowa*, *Oregon* and *Indiana*, with a considerable gap separating the armored cruiser from the battleships. This was reported to Admiral Cervera, and he resolved at once to take advantage of the American absences. His ships, already getting steam up, completed their preparations with great haste. The engine room crew of *Cristobal Colon* had been working on the defense ashore and only now arrived aboard, not having eaten since the previous morning. By oversight there was no food for them aboard ship either, and accordingly their officers dosed them with cognac to keep them going. At 0915 the squadron got under way, proceeding through the channel in the order *Maria Teresa*, *Vizcaya*, *Colon* and *Oguendo*, with 800-yard intervals between ships. The destroyers, with *Pluton* leading, followed 1200 yards astern of the cruisers.

At 0951 the blockaders sighted *Maria Teresa*'s bow as she appeared from behind the cliffs. *Iowa* fired a signal gun and hoisted "Enemy's ships coming out"; *Brooklyn* (Admiral Schley being in tactical command during Admiral Sampson's absence) made "Clear for action" and "Close up". This was in conformity with Admiral Sampson's plan, that the ships should converge upon the entrance and destroy the Spaniards while they were still in the channel. *New York*, already about 12,000 yards eastward down the coast, at once came about and made for the harbor mouth once more, Admiral Sampson (as may well be imagined) greatly agitated and disturbed to be absent from the climactic incident of his command, within sight but out of range.

The blockaders at once closed in, *Iowa* opening fire from 5000 yards; firing rapidly became general. Incredibly, of the American ships only *Oregon* had full steam up; the others (to conserve coal, although there was plenty) had half their boilers lying idle. Indeed, the two armored cruisers fought the entire battle at half power, for their forward engines had been uncoupled, and by a peculiarity of their construction it required twenty minutes at a dead stop to link them up again, an operation for which now there was not the time.

Maria Teresa, already taking hits, slowed at the harbor mouth to drop her civilian pilot (an act of gallant rashness which the other three cruisers emulated), then swung out southwestward and pointed her bow toward *Brooklyn*. She thus constituted a shield for her consorts, and *Vizcaya* and *Colon* cleared the entrance and started down the coast with little damage, largely because she masked them from the American gunners. *Vizcaya* received a single hit abaft the funnels, but otherwise the two were unhurt. *Teresa* fired her forward 11-inch gun at *Brooklyn*, then a full broadside at *Indiana*; her gunners continued to fire as fast as they could, the Americans doing likewise. The gun- and coal-smoke were already so dense that Captain Philips of *Texas* stated: "I might as well have had a blanket wrapped around my head".

Brooklyn, perforce closing the harbor mouth from the west, rather than from the east as her consorts were, quickly found herself closing rapidly with *Maria Teresa*, head-on on opposite tacks. The initial plan of annihilating the enemy at the harbor mouth had already failed due to the distance of the blockaders, and it was clear that a stern chase would ensue. The battleships therefore began to swing to the west to follow *Maria Teresa*; it was clear that *Brooklyn* would have to come about to conform to the general movement, and she chose to turn away from *Teresa* (to port) rather than toward her. She fired with her after turret upon the Spaniard, and *Teresa* fairly soon swung back and proceeded along the coast, unwilling to pursue *Brooklyn* at the risk of being rammed by the oncoming battleships. This fate almost overtook *Brooklyn* herself, as she circled contrary to the movements of the rest. A fortuitous break in the smoke enabled *Texas* to sight her as she swept across the bows of the battleship; Captain Philips gave "All back full" and *Texas* shuddered to a halt not more than a ship's-length away from collision. Before she could regain headway, she had lost about three miles, and was effectively out of the chase, although she continued to bring up the rear. Captain Philips made no official mention of this incident even when Admiral Schley was hotly criticized for his performance in the battle.

Teresa meanwhile suffered heavily from the gunfire of her pursuers. A shell devastated the forward turret, another of 8-inch caliber struck through the shield of a 5.5-inch gun and burst, demolishing the piece, and a 12-inch from *Iowa* pierced the deck near the after turret, cutting steam pipes and scalding the crew of the 11-inch gun. In all she received about thirty hits; her fire mains were shot to pieces, and her wooden decking was splintered and ablaze. Now her excellent ventilation proved her doom, for the unreinforced wood decks blazed uncontrollably, and she was red at every port and "roaring like a furnace", for a forced-draught furnace was in effect exactly what she had become. *Teresa* pointed her bow inshore and struck ground at 1015 at exactly the moment her engines seized up, coming to rest at Punta Cabrera, six miles west of Santiago. *Oregon* gave her a last broadside in passing as she lay already aground, and eventually her survivors were taken off by boats from the armed yacht *Gloucester*.

Vizcaya and *Colon* had, as already mentioned, taken advantage of the flagship's destruction to slip away inshore of her, and were already a distance ahead. *Oguendo*, however, cleared the harbor when *Teresa* was already away to westward, and in consequence was at once engulfed in a vortex of shellfire from the battleships' secondary batteries. One 8-inch shell penetrated the forward turret, demolishing the 11-inch gun and killing the crew; three others passed entirely through the unarmored broadside without exploding. Other shells splintered the decking and set her hopelessly afire, and *Oguendo* turned for shore at 1020. She raised the white flag at 1024, and at 1035 ran ashore so hard that her bow was driven up onto the rocks and she broke her back amidships, coming to rest a half-mile to the westward of *Teresa*.

Meanwhile the destroyers had also come out, with *Pluton* in the lead. By running to the east, or straight out to sea, they might have escaped, but instead they followed in the wake of the cruisers, "seeming determined to expose themselves to the hottest possible fire for the greatest period of time". *Texas*, *Iowa* and *Oregon* all took a shot at them in passing, so to speak; *Indiana*, with her poor boilers, was so far behind already that she slowed to finish them off. The armed yacht *Gloucester*, although greatly inferior in armament and performance to the destroyers, also stood in gallantly to engage them. *Indiana* almost engaged her in error being forestalled by the executive officer's verbal "Look out for the *Gloucester*." *Pluton* received a single medium-caliber shell amidships and, running on the rocks, exploded in a sheet of flame and a cloud of black smoke with the loss of all her crew. *Furor* closed in upon *Gloucester*; the explosive shells from her 1-pounder Maxim "pom-poms" could be seen "walking closer and closer. As they came within 20 yards of the yacht, firing ceased, and it could be seen that *Furor* was circling out of control. It is not possible to say who in all that frenzy of gunfire scored upon her, or with what caliber shells. *Gloucester* closed in firing; and when *New York* came up to add her 8-inch shells, the Spaniard hauled down her flag. She continued to circle and after a brief period exploded and sank; nine of her crew were saved.

Colon was drawing ahead of *Vizcaya*, and the Americans were catching up. *Brooklyn* and *Oregon* led the chase, exchanging fire continuously with *Vizcaya*; *Iowa* and *Oregon* were well astern and being left further behind. *Indiana* was attempting to join in the pursuit, but was left hopelessly far back. Chief Yeoman Ellis, stepping onto *Brooklyn*'s open bridge to confirm the range, was neatly decapitated by a Spanish shell, the only American fatality of the battle. *Brooklyn* had by now drawn up abreast of *Vizcava*, and the Spaniard was afire and being hit continuously. Nevertheless she turned briefly seaward and attempted to ram *Brooklyn*. The American, her eyes already on *Colon*, held her course and outdistanced *Vizcaya*, firing back at her. *Oregon* never ceased to hammer her as well, and shortly she turned back and at 1105 pointed her bow toward the shore. At 1115 she struck a sand bar; her survivors, driven over the side by flames and clustered on the bar; water up to their chests, were fired upon by insurgent soldiers who happened to be camped ashore at this point. *Iowa*'s boats went in to the rescue, the insurgent Cubans having the decency to cease fire rather than endangering the boats of their allies. At length, *Vizcaya*'s magazines detonated in two thunderous explosions, throwing a cloud of smoke 1000 feet into the air. Captain Philips, commanding *Texas,* is quoted as saying: "Don't cheer, boys. The poor devils are dying." Philips also led a prayer of thanksgiving on *Texas* after the battle.

Colon now had a lead of about 10,000 yards, and the Americans settled down to a long chase. She was out of range, and as firing ceased, the ships ran out of the smoke and visibility improved dramatically. Seeing *Indiana* tagging along, Schley made to her "Return to guard harbor entrance" at 1127, for *Reina Mercedes* was still within. It was necessary to repeat the command at 1156 before she complied. *Iowa* was already out of the chase, attempting to rescue survivors of *Vizcaya*, work in which the torpedo boat *Ericsson* also participated. The pursuit thus devolved upon *Brooklyn* and *Oregon*, with *Texas* a distant third and the following *New York* hopelessly out of range astern. Thanks to the enthusiasm of her stokers, *Oregon* far exceeded her designed speed, and even edged *Brooklyn* out of the lead as the chase wore on.

By about 1230 it could be seen that they were definitely gaining upon *Colon*. She had run through her first-rate coal, and had started burning that of inferior quality, and her speed began to fall off a little. *Oregon* fired occasional ranging shots with her 13-inch guns; the first four fell short, but the fifth, fired at a range of 9500 yards, threw up a column of water beyond the fleeing cruiser; the Spaniards knew they were within range. Furthermore, the coast at this point curves southward, and *Colon* had steered along the shore while *Brooklyn* steered direct for the southernmost point of land, thus rendering an eventual interception unavoidable. Thirdly, the stimulant properties of cognac had been exhausted by the stokers, among whom exhilaration was at last giving way to stupor. Thus, at 1315, virtually undamaged, *Colon* hauled down her flag and ran ashore with her seacocks open near the mouth of the Tarquino River, 54 miles west of Santiago. A party from *Brooklyn* boarded and attempted to save her, but despite their best efforts *Cristobal Colon* rolled over to starboard and sank in shallow water at 2300, coming to rest with her port side above water. (*Weapons and Warfare* asserts that *Colon* "was shot to pieces by the *Brooklyn*, and sank . . . with heavy loss of life," but this is false. She surrendered intact, and her entire crew survived.)

Spanish casualties are given as 323 killed, 151 wounded and about 1800 prisoners. It is often remarked, however, that these figures are not definite and Cervera himself estimated his dead at 600 men. US losses were one man dead and one wounded. Both casualties were aboard *Brooklyn*, which ship received 20 shell hits; *Iowa* was hit twice, and *Texas* three times. Out of a total of 9433 shells of all calibers expended, the Americans secured 123 hits. Two 12-inch shells registered, and none at all of 13-inch caliber.

During the evening, a warship came up; at first she was taken for a Spaniard, but proved to be the Austrian armored cruiser *Kaiserin und Konigin Maria Theresa*. So closely did she resemble the *Vizcaya*s that it is a tribute to the steadiness of the American gunners that she was not engaged. One of her officers duly came across by boat to inquire concerning the day's action. He was informed that Cervera's squadron had been annihilated without loss to the Americans, and remarked, "So you have destroyed those four great ships, which I would not have thought possible; and without the loss of a ship or a man. Sir, it is unheard of. I must go to inform my captain." Although as we know one man was in fact lost, these comments still constitute a reasonable post mortem.

The squadron returned to its blockade stations, for *Reina Mercedes* was still inside, and they could not know of her dilapidation, or that her guns were almost all ashore. In fact, the authorities at Santiago had resolved to scuttle her in the channel, to prevent the entrance of the US ships. On the night of July 4 she set on this mission; at 2350 she was observed by the Americans as she entered the searchlight beams, and so heavy a fire was directed against her that her crew abandoned ship, and she drifted clear and sank well out of the way. The city itself capitulated in due course, without any further naval action.

CONCLUSIONS REGARDING THE SPANISH-AMERICAN WAR

It is remarkable that, in the wake of their stunning victories, the officers of the US Navy did not interpret them as a vindication of all their theory and practice. With considerable moral courage, the various officers assigned to analyze the actions came to the conclusion that they had won, not because they were particularly proficient, but simply because they were less clumsy than the Spaniards. In particular, gunnery was abysmal; it will be recalled that against stationary targets Dewey's gunners had secured only 242 hits out of 5859 shells expended, and against ships underway at Santiago the record was still worse, 123 hits from 9433 rounds fired.

It was largely as a result of these figures that Commander W S Sims was encouraged to complete his work on a central fire-control system, which eventually provided the solution to acceptable accuracy, not only in the US Navy but throughout the world. Although his system had been in all essentials perfected in time, such is naval conservatism and inertia that the Russo-Japanese War was fought without it, and it awaited World War I to prove itself in action.

While beyond question the Spanish ships were older, less heavily armed, and in worse repair than the Americans they had to face, we must, I think, look beyond this to explain their total annihilation. While not lacking courage, the Spanish commanders operated in an atmosphere of such utter and pessimistic fatalism, that any initiative was impossible for them. Montojo might have gathered his seaworthy units and attempted to intercept Dewey in transit; instead he chose to ride at anchor and passively await his fate. Cervera might have run straight out for the open sea, precipitating a melee in which some of his ships might have escaped, and in which almost certainly the Americans would have suffered more loss than they did. It was feared at the time, at least in the US press, that he would attempt a raid on the Atlantic coast, but instead he was certain that his command was doomed, and thought only of saving the lives of his men. Confidence in victory is often misplaced, but confidence in defeat is almost invariably self-fulfilling.

In terms of technical and tactical doctrine, these two battles provided nothing worthy of mention. They were both extremely special cases, unlikely to be duplicated, and were so recognized even at the time. The armored cruiser/battleship controversy was no nearer a solution, and a well-armored cruiser was still universally regarded as fit to stand in the line of battle

The fact that in both battles the US forces annihilated the enemy with virtually no loss is entirely extraordinary. It is not to be wondered that contemporary Americans saw in it nothing less than the seal of divine approval upon their imperialist designs.

THE RUSSO-JAPANESE WAR, 1904-1905

INTRODUCTION

(It may be helpful to consult the map in Appendix C, page 266)

Both the Russian Empire and Imperial Japan were pursuing an expansionist policy at this time. Russia, in particular, wanted to obtain a warm-water port as a base in the Pacific since her only deep-water port there was Vladivostok, ice bound for several months in the winter. Her Pacific Fleet was currently stationed at Port Arthur that was leased from China, a tenuous and unsatisfactory situation. Japan had triumphed over China in the Sino-Japanese War and now dominated Chinese foreign policy, particularly so after the Boxer Rebellion of 1898 – 1901. She therefore dominated Korea as well, and she saw the European nations, particularly Russia, as a direct threat to her own expansion in Asia. Conflict seemed inevitable as it proved in fact to be. The general opinion was that Japan could not triumph over a European power, even Russia, as she had against China.

It will also aid in the understanding of strategy to briefly discuss the constraints and circumstances imposed upon the fleet commanders. Total Russian naval forces greatly outnumbered their Japanese counterparts and the quality of Japan's vessels and training of her seamen was only partial in its compensation. The dominant factor in the naval war was that the Russian forces were not concentrated in one fleet either in the Pacific theater, where small groups were based in ports other than the main base at Port Arthur and of course the main reinforcing fleet, the Baltic or Second Pacific Squadron, was in the Baltic Sea and would require months to arrive at the area of the conflict. Both antagonists conducted themselves with this foremost in their strategic thinking and action.

The Russian commanders wished to avoid risking their Pacific Fleet by engaging the Japanese until the Baltic Fleet had arrived. They also had an unjustified confidence in the ability of their army to defend Port Arthur from Japanese land assault. The effort was therefore made to preserve the First Pacific Squadron from attack by the Imperial Japanese Navy even to the point of immobilizing it in Port Arthur. So bizarre were some of the actions taken by Russian officers that suggestions of treason and sabotage persist to this day.

Likewise Admiral Togo, commanding the Imperial Japanese naval forces, was well aware that he would have to engage the Baltic Fleet when it finally arrived. If he did not eliminate the Port Arthur Fleet, or if he did so with a significant loss to his own forces, he was unlikely to be victorious against the Baltic Fleet. Thus, both sides wished to avoid losses before the climactic battle of the war. That Japan succeeded while Russia failed, is what the war was ultimately about. As we shall see, Japan had been preparing for this war over several years, while Russia failed to anticipate how close war was and that Japan would take the initiative when it came. Russia did not concentrate her pacific fleet, did not move it to a more defensible harbor and did not begin preparations for the Baltic fleet to reinforce it until the war was underway. Training and maintenance in the Russian Navy was deficient as well. This was the real failure that predicted the outcome with the consequent rise of Imperial Japan and the ultimate collapse of Tsarist Russia.

It is helpful at this point to digress briefly into the sphere of geopolitics. This war was fought between Japan and Russia, but the battles were fought, land and sea, in China (Manchuria) at Port Arthur, and Korea at Inchon, among other locations, but neither China nor Korea was involved in the struggle. This was a period of colonial expansion at its most blatant. Port Arthur, like Hong Kong and Macau, was leased to a European power. Neither the combatants nor the rest of the world's nations saw any inconsistency in this situation. It is little wonder that China and Korea today feel they must possess weapons in order to deal successfully with other nations. Today those weapons are nuclear.

THE SHIPS OF THE RUSSO-JAPANESE WAR

It will be observed that guns are no longer specifically designated quick-firing; this is because the light and medium-caliber quick-firer had been universally accepted, and it may be assumed that any gun of 6-inch or smaller caliber was so constructed. Also, with the introduction of all-round loading and electrical training and firing for large-caliber pieces, the rates of fire of the two types were no longer as widely discrepant as they had been.

In fact *Kasagi*, *Chitose* and *Takasago* were armed with 8-inch Elswick QF guns. This stretched the quick-firer concept to the maximum, however, and the massive construction required, meant that their rate of fire was not greatly superior to that of standard 8-inch guns; the minor advantage was cancelled out by the awkwardness of the pieces. These guns were therefore not regarded as successful, and the type was not developed. During the 1920's, 8-inch cartridge rounds, as opposed to bag charges, were universally adopted, but these guns possessed a standard breech mechanism rather than the sliding block of the true quick-firer. The largest caliber of naval ordnance to be equipped with a cartridge round was the German 11-inch, but the clumsiness of this gigantic round, and the hazard of numerous huge shell casings rolling about the deck during action, combined to discourage foreign emulation.

Radio communication made its action debut during the Russo-Japanese War, but it must be stressed that this was not the voice radio (or TBS) of World War II origin with which the reader is likely familiar. It was "wireless telegraphy", and we shall refer to it throughout as "wireless" rather than radio, to reinforce our awareness of this distinction. Wireless transmissions were in Morse code; thus communication, although greatly facilitated, was still far from instantaneous. Coding, transmission, reception, decoding and presentation would occupy at least fifteen minutes. Limited range and spotty reception would often require intermediate units to pass the message along to its intended recipient, with all the possibilities for error and garbling this entailed. An hour or more might easily pass between the initial transmission of a message, and its final presentation to its intended recipient in readable form, longer than signal flags.

It should also be pointed out that while the concept of a central fire control was available to navies at this time, neither of the adversaries employed this innovation.   Guns were still trained individually. These and other advances would become standard with the advent of the *Dreadnought* as would the steam turbine and eventually oil-fired boilers to power these ships before the next significant encounter of battleships would take place.

RUSSIAN SHIPS – FIRST PACIFIC SQUADRON

Tsarevitch was a French-built ship completed in 1905. While powerfully armed, she was an unstable ship; her high secondary turrets and towering super-structure constituted an excessive top weight. Further, her exaggerated tumblehome aggravated the tendency to roll, and the tertiary gun ports were set so low that flooding would commence immediately if she took even a minor list. She was armed with four 12-inch/40 guns mounted two each in turrets fore and aft and twelve 6-inch/45 mounted in twin turrets along her broadside. Her thickest armor was 10 inches, and she carried 782 men.

Retvizan was completed in 1901 in the United States. Four of her 6-inch guns were in upper-deck casemates and eight in a main-deck battery. Her main battery was four 12-inch guns mounted in two turrets fore and aft. She was the best Russian battleship involved in the war, and it is interesting to speculate what might have been the results had the five *Orel*-class ships been built to her design rather than to the French *Tsarevitch*'s. Her thickest armor was 10 inches, and her crew comprised 758 men.

Peresviet, completing in 1901, and *Pobieda*, 1902, were Russian-designed and Russian-built. "A thoroughly unsatisfactory design, both armament and protection being poor and speed not exceptional," Conway quite rightly states. Armament was two twin 10-inch/45 guns mounted in turrets fore and aft. These weapons ultimately proved to be of a most unsatisfactory design, requiring alteration and reduction of the propellant charge to be operated safely. Secondary armament consisted of eleven 6-inch/45 guns mounted singly. The massive silhouette made them excellent targets. Their armor was 9 inches at its thickest, and they carried 752 men.

The *Petropavlovsk*'s were Russian-built, completing in 1899. While French influence is obvious, particularly in the secondary turrets, they were more stable than contemporary French-built battleships. Armament was four 12-inch guns mounted in twin turrets and twelve 6-inch. Their thickest armor was 16 inches, their complement 652.

Rossia, like the other large Russian armored cruisers, was a retrograde and a thoroughly infelicitous design, completed in 1897. To all intents and purposes she was a broadside-battery ship, incorporating some modern improvements, Armament was four 8-inch/45 and sixteen 6-inch/45 all mounted in broadside with the 8-inch/45 in embrasures at the ends of the superstructure on either side. It was an outmoded and inefficient configuration, "only half her armament available on the broadside". Thickest armor was 12 inches, with a crew of 842 men.

Gromoboi, completed in 1900 was not so much an improved *Rossia*, but merely a rearranged one. Her armament and armor was similar to *Rossia*, and she carried 877 men.

Rurik was the oldest of these bizarre three-masted broadside armored cruisers completed 1895. *Rossia* was an "improved" *Rurik*, although it is difficult to resist the conclusion that the design should have been retired instead. *Rurik*'s armament was the same as the later versions with only minor variation. Her thickest armor was 10 inches, and she carried 817 men.

After the preceding three monstrosities, it is a pleasure to come upon the trim and orthodox armored cruiser *Bayan*. She was French-built, completing in 1903, but she was a thoroughly conventional design, unlike the French battleships of the period. Her armament was two 8-inch/45 mounted in single turrets fore and aft and eight 6-inch/45 in broadside. Her thickest armor was 8 inches, and she carried 568 men.

Diana, *Pallada* and *Aurora*, Russian-built, completed in 1902, were also thoroughly conventional-looking ships. Armament was eight 6-inch guns mounted in sponsons along her sides and sixteen 3-inch. Their conning were of 6 inch armor, the thickest plate they carried; the crews were of 571 men. *Aurora* was the scene of one of the earliest incidents of the Russian Revolution and is therefore still preserved as a museum ship in St Petersburg, Russia, today.

Bogsatyr was German-built; as with so many of their foreign purchases, the Russians utilized her design for a class of domestically-constructed sister ships. She completed in 1902, and mounted four of her 6-inch guns in twin turrets fore and aft, an unusual feature in a ship of her size. Her thickest armor (conning tower) was 5.5 inches, and her crew numbered 576.

Variag was US-built, completed in 1900; she was a conventional ship of characteristic American appearance, "in fact . . . a diminution of the *St Louis* class". She was armed with twelve 6-inch and twelve 3-inch guns mounted in broadside. Her conning tower was of 6-inch plate (her thickest) and she carried 580 men.

The German-built *Askold*, completed in 1901, is immediately distinguishable from any other warship by her five tall, evenly spaced funnels. The curious raised gun deck immediately aft of the forecastle is a typical German design feature. Armament consisted of twelve 6-inch and twelve 3-inch guns mounted in broadside. The armor of the conning tower was 6 inches thick, and the crew numbered 576.

Boyarin, built in Denmark to the same specification as *Novik*, *Izumrud* and *Jemtchug*, was a far more orthodox design than any of her three sisters. Completed in 1902, she indeed "resembled a lighter version of the *Bogatyr* class" although she did not possess turrets. The six 4.7-inch Canet were in shielded mounts fore and aft one each with the other four in sponsons. She also carried eight 1.9-inch Hotchkiss QF in casements. Her thickest armor was 5 inches, her crew, 266.

Novik, the German response to the *Boyarin* specification, was "lightly built as a much enlarged destroyer", although her armor was of no lesser thickness than *Boyarin*. She was an exceptionally fast cruiser, completed in 1901, with a crew of 537.

Boigi and *Burni*, sisters to the destroyers present at Tsushima, were Yarrow type boats with "two single deck torpedo tubes and one in the stern which was later removed". Completed in 1902, they were unarmored and carried 69 men.

The *Bez Strashni* class were "two-funnel, two-masted destroyers with the 11-pounder forward and three deck torpedo tubes . . . on the centerline with two between the funnels and one abaft them", completed in 1900, carrying 62 men.

The *Vnimatelno* class resembled in appearance the French Frames class. The 11-pounder "was on the 'conning tower' roof forward with a deck torpedo tube abaft each funnel group". Completed in 1902, they were crewed by 59 men.

Leitenant Burakov was an ex-Chinese destroyer seized during the Boxer Rebellion and assigned to Russia. Built in Germany, "she had two large funnels and two deck torpedo tubes and was easily the fastest of the Russian flotilla in the war with Japan." She completed in 1899 and carried 56 men.

Ryeshitelni et al. were of the *Puilki* class, completed between 1901 and 1905. The crew was of 58 men.

Amur and *Yenisei*, "mining or torpedo transports" as they were originally designated, completed in 1899 and are among the first purpose-built minelayers. After the loss of these two ships at Port Arthur, the Baltic Works at once laid down two more minelayers to the identical design naming them in due course *Amur* and *Yenisei*. This duplication tends to confuse the inadequately informed.

The remaining ships of the Russian Pacific Fleet are very much in the nature of antiques and relics, but they were present in the besieged port and so must be listed. An indication of their lack of value is the fact that, apart from the torpedo gunboats *Vsadnik* and *Gaidamak*, even the frugal Japanese did not think it worthwhile to salvage or utilize any of the others.

Of the *Grernyashchis* Conway states, "This class comprised an unusual type of small armored ship. The 9-inch gun was forward and covered in all round, with a large port allowing about 100° training, and the 6-inch aft in a shield." It will be evident that their combat usefulness was extremely marginal. Completed between 1893 and 1894, they carried armor which was 5 inches at its thickest, and their complement was 138 men.

Dzigit completed in 1877, *Razboinik* in 1880; they were three-masted screw sloops of an utterly obsolete type, and their 185 man crews might far better have been employed at some other duty. In some sources they are referred to as "screw clippers" and were intended to function as commerce raiders.

Korietz completed in 1887; "the 8~inch guns were in forward sponsons with 0.75 inch shields, 6~inch aft, the 4.2-inch broadside sponsons." She carried 179 men.

Sivuch and *Bobr* date from 1885. "The 9-inch was mounted as a bow chaser with a 3 pounder training on either side and the 6-inch right aft with a total arc of 270°. Complement was 170.

Vsadnik and *Gaidamak* were German-built, completed in 1890. "One torpedo tube was fixed in the bows and one training on deck." They were unarmored and carried 65 men.

RUSSIAN SHIPS - THE SECOND PACIFIC SQUADRON

While the Russian Pacific Squadron, which we have discussed, was something of a patchwork fleet, the Second Pacific Squadron (comprising virtually everything in the Baltic that could take the sea and fire its guns) was truly as diverse a collection of warships as could well be imagined.

The *Borodino*s were brand-new battleships. Built in Russia with French technical assistance, they "generally resembled the *Tsarevitch*, with the armoring altered, but not for the better." They retained her excessive top weight and tumblehome (a design feature that narrows the hull above the waterline making the main deck beam smaller than lower parts of the hull), and her low tertiary gun ports, and hence shared her tendency to capsize. Their armament consisted of the same four 12-inch guns in twin turrets with twelve 6-inch mounted in turrets in broadside. The thickest armor was 10 inches, and the crew numbered 855.

Osliabia, completed in 1901, was a sister to the *Peresviet*s, and shared in their dismal list of shortcomings.

Sissoi Veliki, completed in 1896, was a perfectly conventional pre-dreadnought. Her main armament was twin 12-inch/40 in turrets fore and aft with the 6-inch/45 QF in barbettes in the main deck battery. Her thickest armor was of 16 inches of nickel steel (inferior to the 10-inch maximum Krupp cemented armor in the newer battleships), and she carried 586 men.

Navarin was completed in 1896, and was originally classified as a turret ship. "A most unusual feature was the location of the four funnels in two pairs abreast," although she was a distinctive-looking snip anyway, with a curiously antiquated appearance which was accentuated by the location of her secondary battery of six 6-inch/35 in casements on the upper deck. The main battery was a more conventional twin turreted 12-inch/35 fore and aft. Her thickest armor was 16-inch compound, her complement 622 men.

It is not possible to find a single adequate illustration of *Imperato Nikolai I*. Her near-sister *Alexander* features a bizarre hooded barbette for her twin 12-inch/30 instead of a turret like *Nikolai*'s. The

151

Japanese refitted her after her capture. She was cut down extensively aft, altering her appearance substantially from what it was during her Russian service. She completed in 1891; originally classed as a Turret Ship, she was really not fully fit to stand in the line. Her thickest armor was 14-inch compound, and she carried 611 men.

Ushakov, *Seniavin* and *Apraksin* were originally designed for Baltic coast defense, specifically intended to counter the three Swedish *Svea*-class ships. To send them around the world to engage the first-class Japanese Navy was the height of irresponsible folly. (A number of sources attribute 10-inch guns to all three, but it seems on balance that *Ushakov* and *Seniavin* in fact carried 9-inch weapons. *Apraksin* was equipped with a twin forward turret, with a single turret aft. The thickest armor was 10-inch Harvey, the complement 404.

(The armored cruisers of the Second Pacific Squadron were all obsolete, the modern ships having been stationed in the East from before the war.)

Admiral Nakhimov, completed in 1888, was like a Frenchman, although in fact she was a copy of the British *Warspite*-class armored cruisers, "of which the Russians had managed to obtain drawings, although they contrived, by alteration of bunkers and hatches, to impair the protection of the machinery." By 1905 her rig had been much reduced, being cut down to two short military masts. The 8-inch guns were in four twin turrets, two fore and aft and the remaining two sponsoned amidships. The thickest armor was 10-inch compound, the crew numbered 570.

Although not officially classed as sisters, *Monomakh* and *Donskoi* were virtually identical in appearance. They were ancient broadside ironclads completed in 1885; both, however, had been extensively modernized and completely rearmed, *Donskoi* in 1895 and *Monomakh* in 1898. Notwithstanding, neither ship was by any stretch of the imagination fit to stand against the Japanese Navy. Their armament consisted of four 8-inch sponsoned two on either side and twelve 6-inch guns between these and on her ends outside the battery. The thickest armor was 6-inch compound in both; the complement of *Monomakh* was 495 men, and that of *Donskoi*, 507.

Diana and *Aurora* have been discussed above and *Oleg* was a sister to *Bogatyr* also discussed above.

Svetlana was a trim little protected cruiser, completed in 1897. She was fast, but since in peacetime she had been the yacht of the Grand Duke commanding the Navy, she was fitted out with an excessive amount of woodwork, which could not but impair her ability to sustain shellfire. She was equipped with six 6-inch guns and two 15-inch torpedo tubes. Her thickest armor was 5 inches, and she carried 101 men.

Izumrud and *Jemtchug* were brand new, Russian built ships, copies of the German-built *Novik*, save with three masts. Armament was six 4.7-inch guns and four 18-inch torpedo tubes. Armor was 3 inches at its thickest and 550 men comprised the crew.

The destroyers were all of the *Boiki* class. They carried one 11-pounder and five 3-pounder guns and three 15-inch torpedo tubes and 18 mines.

JAPANESE SHIPS

In addition to the ships listed below, the Japanese Imperial Navy still possessed virtually the entire squadron listed in the SHIPS OF THE SINO-JAPANESE WAR, as well as the Chinese ships captured at the capitulation of Weiheiwei. These obsolete tubs were already very much in the category of museum pieces, with the exception of the protected cruiser *Yoshino*.

The Japanese Navy, unlike the Russian "collections of samples", possessed in its heavy units a truly unified battle fleet. All its battleships and armored cruisers being built to similar specifications with similar firepower and performance. (While this is largely a function of the fact that almost all the ships were British-built, the foresight and right-thinking of the Japanese command cannot be dismissed.)

The "near-sisters" *Shikishima*, *Hatsuse*, *Asahi* and *Mikasa* are sometimes regarded as one class, and sometimes divided into the two classes *Shikishima-Hatsuse* and *Asahi-Mikasa* on the basis of appearance. In fact the four are progressive improvements upon a single design, which was a copy of that of the British *Majestic* class. The vessels carried a maximum of 14 inches of armor; complements were as follows: *Shikishima* 856, *Hatsuse* 741, *Asahi* 856 and *Mikasa* 830. The vessels completed, in the order given above, between 1900 and 1902. The ships were a great improvement on *Fuji* with much improved armor protection and more powerful armament. Speed and also the caliber of the armament were similar to *Fuji*", so that the six battleships "were able to operate together as a homogenous unit."

Fuji and *Yashima* were built to the design of the British *Royal Sovereign* class, except with smaller caliber main guns; they completed in 1897. The guns, of a newer design, were as powerful as the *Royal Sovereign*'s 13.5-inch weapons as well as being enough lighter that they could be protected by 6-inch thick turrets. The 12-inch guns could only load when trained along the line of bearing; thus their rate of fire was significantly lower than that of the *Shikishima*'s with their all-around loading. These six were also equipped with between ten and fourteen 6-inch QF guns. The compound armor's maximum thickness was 18 inches in *Fuji*. This was in fact inferior to the *Shikishima*'s

Harvey steel armor, which provided 33% greater protection than did compound armor of the same thickness.

At the same time that the *Shikishima*'s were ordered, a specification for six armored cruisers was also promulgated. *Asama, Tokiwa, and Idzuno, Iwate* were built in Britain, *Yakumo* in Germany and *Adzuma* in France, all completing in 1899 and 1900. The *Idzumo*s differed from the *Asamas* in being provided with a new and more efficient type of boiler. They were armed with four 8-inch/45 mounted in fore and aft turrets and twelve or fourteen 6-inch QF. Beyond this, the ships differed only insofar as the ship building styles and capacities of their builders did. Thickest armor was 6 inches; the *Idzumo*s carried 672 men, the *Asama* 726, *Yakumo* 700 and *Adzuma* 650.

The Kasuga's were of Italian construction, a year old, near-sisters to the Spanish *Cristobal Colon*, and were acquired through Argentina. *Kasuga* mounted the 10-inch gun forward, with twin 8-inch aft; *Nisshin* carried two twin 8-inch turrets. The thickest armor was 6 inches, the complement 600.

Kasagi, Chinose and *Takasago*, completed in 1898, were of similar design and externally almost indistinguishable. They were designed by Sir Philip Watts and are typical Elswick cruisers. The 8-inch/45 QF gun was not an unqualified success. These two weapons were mounted fore and aft behind gun shields. They also carried ten 4.7-inch QF. The thickest armor was 4.5 inches; *Takasago* carried 425 men, *Chinose* 454 and *Kasagi* 405.

Tsushima and *Niitaka* were brand-new ships, built in Japan to a domestic design, seaworthy and powerful vessels which outclassed many other contemporary protected cruisers. Main armament was six 6-inch/40 guns. Thickest armor was 4 inches, compliment 320.

Otowa was a smaller, faster version of the *Tsushima* class, but with a lighter armament. Her age was a little less even than the *Tsushima*s, her armor 4 inches at thickest, and her complement 512.

Suma and *Akashi* were the first cruisers to be built entirely according to a Japanese design with Japanese materials, except for the guns which were imported from Britain. The design and disposition of the armament were similar in many ways to *Akitsushima*, and dimensions were practically the same. *Suma* completed in 1896, *Akashi* in 1899. Armament was two 6-inch/40 QF and six 4.7-inch. Heaviest armor was the 4.5 inch-thick gun shields, and the complement was 510.

Izuma, ex-*Esmeralda*, was purchased from Chile in November 1894, but proved unstable in Japanese waters, having been designed for the relatively tranquil South American coastal waters. Her armament of two 10-inch BL and six 6-inch BL guns was replaced in 1902 with two British 6 inch BL Mark VII and six 4.7-inch QF guns. The 6-inch BL Mark VII was capable of rates of fire equal to 6 inch QF and the propellant charge in silk bags as opposed to brass cartridges saved weight and space in the magazine and eliminated empty brass shell casings rolling loose about the guns. The Mark VII and slightly modified Mark VIII were very successful designs and saw service into World War II. *Izuma*'s thickest armor was only one inch however; this plus her age kept her from any significant action during the war, relegating her to patrolling shipping lanes and other such duties. She was present at the Battle of *Tsushima* however and did exchange fire briefly with the Russian *Vladimir Monomakh*. By 1904 she was virtually obsolete at the age of twenty years. Her crew numbered 300, One interesting sidelight was that Chili arranged to sell *Esmeralda* to Ecuador who then immediately resold her to Japan in order to maintain the neutral status of Chili. *Esmeralda* flew the flag of Ecuador during her voyage from the Galapagos Islands, Ecuador, to Japan thus technically serving in three different navies.

The ships described in the INTRODUCTION TO THE SHIPS OF THE SINO-JAPANESE WAR were almost all still in service. *Naniwa* and *Takachiho* had been re-armed with eight 6-inch Elswick QF guns in place of the 10.5 and 5.9-inch, but otherwise the vessels were not much altered. Certain captured Chinese ships were also in service, as described in the NAVAL BATTLES OF THE SINO-JAPANESE WAR. These ships were for the most part confined to scouting and fire-support missions.

The *Ashagiri*s were "built in Japan but developed from the Thornycroft-designed *Shirakumo* class," dating from 1905. Complement of these boats is not a matter of record.

Akatsuki and *Kasumi* were Yarrow built (British) boats, completing in the winter of 1901-1902 carrying 59 men. *Asashio* and

Shirakumo were completed by Thornycroft (British) in 1902 and carried 59 men.

The *Akebono*s were Yarrow boats, with the 12-pounder mounted aft, but apart from their lighter armament much similar to the *Ashagiri*s. They completed in 1899, and their complement is not recorded.

The *Kagero*s with their aft~mounted 12-pounder were Thornycroft boats, completed 1898-1906 and carrying 54 men.

Torpedo boats built subsequent to the Sino-Japanese War were obsolete when completed. Classes were: Second-class *TB67-TB75* (Japanese-built); Third-class *TB50-TB59* (Japanese); Second-class *TB39-TB43* and *TB62~TB66* (Brit1sh); and second-class *TB5I-TB6l* (German).

NAVAL OPERATIONS OF
THE RUSSO – JAPANESE WAR, 1904 - 1905

(It may be helpful to consult the map contained in Appendix C, p. 266)

INITIAL OPERATIONS

ATTACK ON PORT ARTHUR, FEBRUARY 8-9, 1904

The diplomatic situation between Russia and Japan had deteriorated to such an extent that it was obvious to all sensible observers that war would come within days. The squadron at Port Arthur, comprising *Petropavlovsk* (flag of Admiral Stark), *Sevastopol*, *Poltava*, *Peresviet* (flag of Rear-Admiral Prince Ukhtomski), *Pobieda*, *Tsarevitch*, *Retvizan*, *Bayan*, *Askold*, *Diana*, *Pallada*, *Egvik*, *Boyarin* and destroyers and light craft, had been placed on alert. The ships were still in their peacetime colors, white with black funnels; but torpedo nets had been ordered out (though not all ships complied). *Askold* and *Pallada* illuminated the area with searchlights, and the destroyers *Rastoropni* and *Bez Strashni* were dispatched around dusk of February 8, to make a twenty-mile sweep of the approaches. The ships, arrayed in the roadstead outside the harbor proper, were still lighted, and navigational beacons were lighted and in place. It was a raw, misty night, with light snow gusting over the sea.

Admiral Togo arrived with the greater portion of the Japanese Combined Fleet in the environs of Port Arthur on the evening of February 8. His flag was in *Mikasa*, with *Asahi*, *Hatsuse*, *Shikishima*, *Fuji*, *Yashima*, *Idzumo*, *Iwate*, *Tokiwa*, *Yakumo*, *Adzuma*, *Kasagi*, *Chitose*, *Takasago*, *Yoshino* and destroyers in company. Promptly 10 destroyers were sent in to attack the Russian ships, although a declaration of war had yet to be uttered. The First Division comprised *Shiragumo*, *Asashio*, *Kasu*mi and *Akatsuki*; the Second, *Ikazuchi*, *Oboro* and *Inazumo*; and the Third, *Usugumo*, *Shinonome* and *Sazanag*i. The destroyers were darkened for the approach save for a single white stern light. The moon had not yet risen, and the Japanese fell into considerable confusion. *Oboro* and *Ikazuchi* were damaged in a collision with one another, throwing the Second Division into disorder, and *Sazanami* approached *Rastoropni* and *Bez Strashni* in error as the Russians passed between the Second and Third Divisions, jeopardizing the surprise of the attack. The four new boats of the First Division were the first to gain visual contact with the Russians squadron, and the first to attack.

The four destroyers came to full speed (a nominal 31 knots) for the run-in; they were illuminated by *Pallada* during the approach, but she mistook them for the returning Russian patrol and took no action. The two Russian destroyers had in fact entered the harbor and were in the process of reporting their suspicious contact when the Japanese fired torpedoes from a range of 650 yards. Surprise was complete when one of *Asashio*'s torpedoes exploded against the side of *Retvixan* beneath the forward turret. A hole of 215 square feet was torn in the hull below the armor belt, and 1000 tons of water poured into the ship. Officers and men were thrown to the deck by the impact of the blast, and *Retvizan*'s searchlights illuminated spontaneously, spinning dizzily out of control.

Almost immediately, a second torpedo struck *Pallada*'s coal bunker amidships; a sheet of fire raced up her side, "dispelling any lingering notions that this was some sort of exercise." Searchlights came on throughout the Russian squadron, and an inaccurate but intense fire was opened. It was too late to prevent one of *Akatsuki*'s torpedoes from striking *Tsarevitch*, aft on the quarter, tearing a hole of 533 square feet in her hull. As she withdrew, *Akatsuki* (last ship of the First Division's line) began to come under aimed Russian fire, although she was not hit.

The ships of the Second and Third Divisions, making the approach in considerable disorder, were received by rapid but largely undirected Russian fire. The Russian searchlights glared into the night, making accurate torpedo aiming difficult. Further, the volume of fire striking the sea induced the Japanese to fire torpedoes at a greater range than had the ships of the First Division. In consequence, no further hits were obtained, although four long yellow torpedoes were later discovered fouled in the torpedo nets of *Tsarevitch*. Eighteen torpedoes had been fired (*Ikazuchi* had dropped out of the attack following the collision) for a total of three hits. The gallantry of the crew of *Oboro*, which pressed home her attack despite collision damage which rendered her only partially controllable, is worthy of specific mention.

In all three Russian ships, damage control quickly isolated the flooded compartments. *Pallada's* fire had been spontaneously extinguished by the inrush of water, and neither of the two battleships took flame. An ill-advised attempt was made, as soon as the battleships could get underway to bring them into the security of the inner harbor. In consequence of the water they had shipped, however, they were riding so deep that both grounded in the narrow channel. Thus the undamaged ships were trapped in the open roadstead, vulnerable to any further attack the Japanese might make. Of all the Russian ships, only the little cruiser *Novik* was able to raise steam in time to pursue the Japanese. She failed to make contact, and had rejoined the squadron by 0500.

At dawn, a reconnaissance force consisting of the protected cruisers *Chitose*, *Kasagi*, *Takasago* and *Yoshino* came to within sight of the anchorage at Port Arthur. Admiral Dewa, observing the Russian ships still in the outer roadstead and disposed "in no sort of order", judged that the situation called for an immediate attack utilizing all available units. He advised that "the greatest results" might reward prompt action.

In accordance with this report, the entire Japanese force at once made for the anchorage, coming into visual contact with the Russian ships at around 1200. The day was clear, with a scant haze along the coast; the sea was smooth under a slight southerly breeze. The battleships stood in line ahead in the order *Mikasa*, *Asahi*, *Fuji*, *Yashima*, *Shikishina*, *Hatsuse*; the armored cruisers formed the Second Division; Dewa's protected cruisers the Third.

The Japanese opened fire at 8500 yards. The Russian ships, maneuvering evasively under cover of the forts, returned the fire at once, as did the guns ashore. The Russian fire was immediately effective; *Mikasa* received three shell hits, one of 10-inch caliber, which wounded seven men; *Shikishima* was struck by a 6-inch shell that burst against the forward stack, raking the upper decks with splinters and wounding seventeen men; *Fuji* and *Hatsuse* were each hit twice, with two dead and seven wounded, and seven dead and nine wounded respectively. During the firing pass of the Second Division,

Novik attempted a torpedo attack, but was struck amidships by an 8-inch shell from *Yakumo* and retired without discharging torpedoes. *Askold*, *Diana* and *Bayan* were likewise damaged by shell fire, as were several of the Japanese cruisers.

After a single long-range firing pass, the Japanese ships withdrew, and firing had ceased by 1245. (In assessing Admiral Togo's caution during this and subsequent actions during the siege of Port Arthur, it must be remembered that he had always to reckon with the Russian Baltic Fleet, that he well knew would eventually be sent to oppose him. Thus he could not accept serious loss, even if the result was the destruction of the Port Arthur Squadron; for the effect would have been to leave Japan partially defenseless against the enemy's back-up force. When at length, at Tsushima, all is at stake; we shall see him in quite another guise.)

ELIMINATION OF RUSSIAN UNITS AT INCHON (CHEMULPO) FEBRUARY 8 - 9, 1904

Russian peacetime dispositions included a subsidiary force on station at Inchon, Korea, consisting of the protected cruiser *Variag* and the ancient gunboat *Korietz*. The presence of this force was known to the Japanese, who maintained the old armored cruiser *Chiyoda* on the same station. While planning the neutralization of the Port Arthur squadron, the Japanese had taken this small contingent into account as well. Its destruction had been assigned to the Fourth Cruiser Division of Admiral Uriu, consisting of *Naniwa*, *Takachiho*, *Niitaka* and *Akashi*, and reinforced for the occasion by the armored cruiser *Asama*. *Chiyoda* was alerted to the impending hostilities and left Inchon by stealth during the night of February 7, joining Uriu's squadron during the following morning.

On the morning of February 8, more than 12 hours before Togo arrived at Port Arthur, *Korietz* was dispatched to Port Arthur carrying mail. Fairly soon she encountered the six approaching Japanese cruisers escorted by three torpedo boats. Comprehending the hostile purpose of this force, *Korietz* at once turned and made her best speed back toward Inchon. The Japanese torpedo boats attempted to head her off, and in fact got into position to fire three torpedoes at her; none of them hit. *Korietz* fired a few rounds at the torpedo boats, and regained the harbor of Inchon without damage. The Japanese resolved not to attack the harbor at once, since a number of neutral warships were also present, and it was feared that international repercussions might be considerable, Instead at 0900 on February 9, an ultimatum was presented to the two Russian captains. In essence this was: Sail by 1600 or be sunk in harbor.. At 1100 the Russians began to raise steam and prepared to leave Inchon. They stood out boldly toward the waiting Japanese, and at 1145 *Asama* opened fire with her 8-inch guns at 7500 yards; *Variag* replied at once, and the remainder of the Japanese joined in.

Five hits were scored upon *Variag* in rapid succession. An 8-inch shell destroyed the upper bridge, set fire to the charthouse, toppled the foremast and eliminated the crew of the forward range-taking station. By 1155 *Variag* was afire in many places; her steering engine

162

was knocked out. She zig-zagged by means of hand signals and controls until an 8-inch hit below the waterline caused her to quickly take a list to port. Two of her funnels had fallen, the after bridge had been destroyed, and new fires were breaking out constantly. *Variag* came about and steered back for Inchon. Her mainmast fell and she was badly ablaze, but some of her 6-inch guns continued to fire.

Korietz, unengaged, had fallen astern of *Variag*, but as the crippled cruiser fell back toward Inchon, the gunboat attempted to escape by continuing on course inshore of *Variag*, masked by smoke and shell bursts in the water. Slow old *Chiyoda* had fallen astern of the other Japanese cruisers, however, and she observed *Korietz'* maneuver. Closing, *Chiyoda* opened fire upon her. A single hit from *Chiyoda's* 4.7-inch guns set the ancient gunboat afire, and *Korietz* came about and followed *Variag* back to Inchon.

Both Russian ships regained the port, and the Japanese forbore to pursue them into the restricted waters of the harbor, deterred for the most part by the danger of injury to the foreign warships that were also present. The Russians recognized that their case was hopeless; the crew of *Variag* opened her seacocks, and that of *Korietz* set a slow fuse to her magazines. The seamen then abandoned ship and were taken aboard neutral vessels (the USS *Vicksburg* refusing to receive survivors). Subsequent Japanese attempts to detain the Russians were rebuffed by the foreign naval officers. At 1537 *Korietz* exploded with a prodigious blast and sank at once. *Variag* sank lower and lower in the water, rolling sluggishly, until she finally sank, with only her topmasts and steampipes visible above the surface of the harbor. (*Variag* was subsequently salvaged by the Japanese and repaired. She served as a cadet training ship until March 1916 when she was sold to Russia being delivered on the 5th April and renamed *Variag* once more. Her name during her Japanese service having been *Soya*).

Variag is celebrated in Russia to honor the stoicism and valiant action against a much larger Japanese force in what is referred to as the Battle of Chemulpo Bay in poetry and song. During the Russian Revolution her crew mutinied and refused to sail, whereupon she was seized by Britain and served in yet another navy until she ran aground while under tow and was eventually scrapped in place.

BLOCKADE OF PORT ARTHUR,
FEBRUARY 10 - MAY 25, 1904

As we have seen, Admiral Togo determined that the Port Arthur Squadron (soon to be renamed the First Pacific Fleet) could not be annihilated outright without incurring an unacceptable level of risk. Thus, as a rather belated declaration of war was uttered on February 10, the Japanese fleet began the imposition of a close blockade upon the Russian base.

On February 11, the Russian minelayer *Yenisei* was sent out to mine Talien Bay, as a precaution against the Japanese using it as a forward base. A total of 400 mines had been laid, and, as the wind was picking up and the sea rising, the operation was about to be broken off when a lookout reported that one of the mines had come to the surface. It would have been impossible to re-moor the mine in view of the condition of the sea. Unwilling to leave it adrift (both because this would have been a violation of international law and because of the danger that the enemy might retrieve it and learn its structure and operation), *Yenisei* was maneuvering for a favorable position from which to explode the thing by gunfire when she ran onto another of her own mines. Fatally damaged, she sank in twenty minutes with the loss of 5 officers and 89 men.

As *Yenisei* grew overdue, and the sea continued to rise, it was deduced that she had come to some harm, and *Boyarin* was dispatched to find and succor her. As the position of the newly-laid mines was (naturally) unknown to her, *Boyarin* proceeded some distance to seaward, making a visual search of the coast. Her precautions availed nothing, however, for she too struck a mine. The crew abandoned ship; it was thought that she might be towed next day, but in the course of the night she drifted onto another mine and sank.

Also on February 11, the four cruisers stationed at Vladivostok sortied, encountering and sinking two Japanese merchantmen in the Tsugaru Straits. The Russians returned to port without incident. On February 12, the Japanese once more bombarded the ships outside Port Arthur, without significant results.

Various observations indicated that the Russians had succeeded in circumnavigating their stranded battleships, and that the channels

164

into Port Arthur were again in use. Accordingly, on February 20, Admiral Togo decided to implement a plan to block the channels with redundant merchantmen, painted black and filled with concrete and rocks, which were intended to be piloted into the correct positions and then deliberately sunk by their crews. Japanese sources attribute the origin of this scheme to a 1903 memorandum by Lieutenant-Commander Rvokitsu Arima but it is difficult to imagine that the US Navy's attempt to block Santiago Harbor by sinking the collier *Merrimac* in the channel was not a major source of Arima's inspiration

Volunteers were solicited, and around 2000 responses were received; some aspirants submitted applications written in their own blood to emphasis their sincerity. Seventy-seven were chosen.

The attack took place on the night of February 14 - 15; the sea was calm, the weather fine; the block ships went in five abreast, with escort of twelve destroyers. *Retvizan*, still aground in the channel, sighted the approaching merchantmen, although she did not see the destroyers lurking astern of them. Still, her suspicions raised by the unusual formation of the unidentified steamers, and by the time of night, she illuminated them and opened fire. The enemy ships increased speed and bore in, "rather than turning away with sirens shrieking as our own ships would have done," and the destroyers removed all doubt as they darted out and fired upon *Retvizan*. Firing became general throughout the Russian squadron and forts, and all searchlights focused on the oncoming ships.

One was hit at once and sank at a considerable distance from the port; it is occasionally said that the captain of this ship opened her seacocks prematurely. The others bore on resolutely, the destroyers once more falling into station astern. Two more block ships were disabled fairly rapidly by gunfire, and their survivors taken off by the destroyers, which then withdrew (One of the derelicts eventually ran ashore on the rocks beneath White Wolf Hill, the other coming to rest on the southern slope of Lighthouse Hill.)

The two remaining ships, *Jinsen Maru* and *Hokoku Maru*, continued to close in, undaunted by the increasing Russian fire. The Russian gunners, in the confusion of night action, imagined that they were confronted by a full-dress amphibious assault, and the vigor of their resistance was in proportion. *Jinsen Maru* blew up and sank in the shadow of Golden Hill; the Japanese assert that she struck a mine. *Hokoku Maru* continued on alone, coming so close to her target that *Retvizan*, stranded in the channel she meant to block, believed that she was attempting to ram. In this, of course, she "failed", since it had not been her intention, but as she closed the battleship her unarmored hull was riddled and she went down short of her objective, becoming a nuisance to navigation rather than an obstruction.

With the light of morning, a reconnaissance force, consisting of *Bayan*, *Askold* and *Novik*, proceeded to examine the wrecks and bring off survivors. As students of our own Pacific war will appreciate, this latter was more easily said than done. The Japanese, regarding surrender as the ultimate dishonor, chose suicide instead, and only three men were subdued and brought aboard the Russian ships.

At this point a squadron of Japanese cruisers appeared, followed in short order by the ships of the First and Second Divisions (the battleships and armored cruisers). *Bayan* and her consorts were thus in a position of some peril, being at a distance from the port, as the Japanese opened fire upon them at long range. They replied, as did the forts; under cover of the land guns, the three Russian cruisers ran along the coast for home. They succeeded in regaining harbor, though not without damage, and only after having spent a hot hour or so threading the tortuous channels under fire. The Russian naval losses were 22 men killed and 41 wounded; 21 men ashore were killed or wounded as well.

As the Japanese drew off and were lost to sight, the destroyers *Bez Strashni* and *Vnushitelni* set out to intercept the retiring enemy and attack with torpedoes. (As this was the only offensive maneuver attempted by Admiral Stark, we may as well remark at this point upon his curious "slight-risk" philosophy. The ridiculous parsimony of dispatching two destroyers eight percent of his available destroyer force, in pursuit of the enemy's main battle squadron will be apparent. The fate of the two at the hands of four protected cruisers will point up the probable result, had they succeeded in their overly-ambitious mission.)

In the afternoon haze the Japanese eluded their diminutive pursuers. The two destroyers had abandoned the "chase", and were returning home along the coast, when they were sighted by Admiral Dewa's scouting force, consisting of the protected cruisers *Chitose*,

Kasagi, Takasago and *Yoshino*. The destroyers ran for home, opening fire (one suspects for moral effect, since accuracy must have been minimal) at the extreme range of 11,000 yards. It is not clear whether or not the Japanese returned fire at this time, although they unquestionably took up the pursuit.

Bez Strashni ran straight for Port Arthur, and gained safety without sustaining damage. *Vnushitelni*, marginally slower, ducked into Pigeon Bay, unobserved by the enemy in the gathering darkness. The cruisers, which seem to have lost visual contact almost immediately, nevertheless conducted a search of the coast. By Dewa's testimony, they were about to give up the hunt, rather than risk coming any closer to the Russian shore emplacements, when *Vnushitelni*, thinking herself discovered as the Japanese "closed in" on her hiding place, once more opened fire and tried to run for Port Arthur

The cruisers returned fire and gave chase. Thinking it inevitable that they should overtake her, the captain of *Vnushitelni* set his vessel afire and gave "Abandon ship". As the crew took to the boats, *Yoshino* closed in to finish off the destroyer. Three 6-inch shells struck *Vnushitelni*, these hits, together with her crew's demolition efforts, damaged her beyond repair; some 6-inch shells fell in the vicinity of the boats in which the Russian crew had embarked, although there is no record of casualties. Fear of mines, and the fire of shore batteries, induced *Yoshino* to withdraw. She reported that *Vnushitelni* had "lost all fighting power", and "must have gone aground since she had not sunk." In fact, with the departure of the Japanese her captain returned aboard and opened her valves, and she sank. Her crew then went ashore in the boats and walked back to Port Arthur overnight.

The men of the fleet were eager for action. It seemed to them that the destroyers had made a good showing against overwhelming odds, and they very much wished to try the issue on more even terms. Also, accounts of the stand of *Variag* and *Korietz* at Inchon had begun to circulate among the men of the Port Arthur Squadron, and the heroism of their compatriots in another hopeless battle further fired their own ardor.

Admiral Stark, however, was thrown into a funk by the loss of *Vnushitelni* and his policy, already one of extreme diffidence, became henceforth one of virtual paralysis. The decision had already been made to replace him with Admiral Makarov, who was enroute overland, and it may be that knowledge of his imminent supersession contributed to stay his hand. The fact is that for ten days from February 26, the Japanese did not appear in daylight before Port Arthur, and no Russian attempt was made to seek out the enemy. The ships remained inactive in port, and "morale in the squadron plummeted".

Skilled engineers and dockyard workers had arrived from the West to repair the ships damaged in the surprise torpedo attack. Port Arthur lacked personnel competent to perform the repairs. Indeed, she had not even a dry-dock of sufficient capacity for battleships. A cofferdam (invented as it chanced, by the same Makarov twenty years before) was placed alongside *Retvizan* to permit a patch-up, and on March 7, she was floated and entered the harbor under her own power. On this same day Makarov arrived. Finding Stark's flag still flying in *Petropavlovsk,* he hoisted his own in *Askold*. He chose the cruiser, rather than one of the battleships, for characteristic reasons; she had seen action, and she was fast.

Makarov at once instituted a policy of vigorous night destroyer patrols, to counter the Japanese destroyers that haunted the coastline during the hours of darkness, laying mines, shooting up coastal installations, and generally creating such mischief as they might. On the night of March 8-9 the coastal patrol fell to the Japanese First Destroyer Division, consisting of *Shirakumo, Asaenie, Kasu*mi and *Akatsuki*. In the midst of their all-night patrol, they fell in with a like number of Russian destroyers. (Sadly, the only available account of this action is given by the elusive Warners, who seem to have consulted only Japanese sources in this instance, and who fail to give the names of the Russian boats.) A brisk action ensued in the darkness at ranges down to 50 yards. One Russian torpedo was fired, and a Russian destroyer was reported "fleeing in flames".

On the following night, the Third Destroyer Division (*Akebono, Usugumo, Shinakumo* and *Sazanami*) took their turn at the harbor patrol. The sea was relatively calm, the weather clear and intensely cold. Withdrawing at sunrise after a quiet night, the Japanese sighted *Ryeshitelni* and *Steregushchi*, returning home from a similar mission. Unseen in the early-morning dusk, the Third Division maneuvered to cut off the Russians retreat to Port Arthur before opening fire at

168

300yards.   The outnumbered destroyers showed no disposition to escape, however, and a closely-fought engagement ensued.

The two Russians stood up to their four assailants. Casualties were sustained by every boat present except *Sazanami*. In fairly short order *Ryeshitelni*'s steering linkage was shot away, and she headed for Port Arthur, maneuvering with her engines and commanded by a midshipman, the highest-ranking officer not killed or incapacitated by wounds. *Stereguschi* served as rear-guard for her stricken consort, conducting a fighting retreat.

Ryeshitelni made good her escape, but the gallant *Stereguschi,* the race all but won, succumbed almost within range of the Port Arthur's guns. A Japanese shell hit blew up her engines; the stern was wrecked and the crew of the after gun eliminated. She went dead in the water, but her forward gun remained in action. *Sazanami* was ordered to close in and finish off the cripple. Japanese fire quickly disabled the forward gun, and the Russian crew took to the water. The captain of *Sazanami* has left a cold-blooded memoir, quoted by the Warners, in which he writes:

> I saw a great column of thick steam escaping and some men trying to get away. These we were killing and wounding with deadly aim. It was indeed a sight which is seldom seen at sea, and from which one derives peculiar gratification--fighting against men full of vitality instead of inert steel.

The Japanese boarded the hulk and took her in tow, but two ratings who had barricaded themselves below opened her valves and went to their deaths with her. The Japanese noticed her quickly becoming sluggish and beginning to wallow, and had barely time to get their party away before she parted the tow and went down.

Informed of the action, Admiral Makarov had at once transferred his flag to *Novik*, for with her shallow draft she was the only ship that could proceed immediately to the rescue; for at low tide, as at this time, none of the deeper ships could negotiate the channel. (This is the first mention of tidal difficulties in regard to the use of the port.   It says much of Admiral Stark that the fact that it was an operation requiring precise timing and the better part of twenty-four hours to get the fleet to sea never inconvenienced him at all.)

Ordering *Askold* to follow him immediately she had steam up, and the rest of the fleet to raise steam so that they might emerge without delay as soon as the tide permitted, Makarov set off at once in the unescorted *Novik* to rescue *Stereguschi*. As we have seen, he was too late; when *Novik* arrived on the scene the Russian destroyer was gone, and the Japanese were already retiring. *Novik* set out in pursuit of the four destroyers—only to sight the four scouting cruisers of Admiral Dewa, closely followed by the First and Second Battle Squadrons. Even so aggressive a commander as Admiral Makarov eschewed such odds; *Novik* came about and beat a hasty and successful retreat to Port Arthur.

The arrival of the Japanese battle fleet had no connection with the destroyer skirmish; it was their mission to contain the Russian squadron and prevent it from interfering with a troop landing at Pyongyang. In this it was entirely successful; the landing was effected without loss, and Makarov did not even learn of it until the following day. Admiral Togo made use of the occasion to put into practice a system of indirect firing. The battleships and armored cruisers took up positions such that they could lob shells into Port Arthur over the surrounding mountains, without being visible to the Russian ships or shore gunners. Trapped in the harbor by the state of the tide, the Russian ships endured a six-hour bombardment to which they were utterly unable to reply. Japanese accuracy was minimal, as was to be expected: *Retvizan* sustained some damage, and one of *Askold*'s guns was demolished, but otherwise the Russians escaped harm. Morale was naturally affected by the one-sided shooting, but even this was largely offset by the example of Admiral Makarov, who spent the six hours touring the harbor in his ceremonial barge seemingly oblivious to the towering columns of water raised by the 8 and 12-inch shells.

As soon as the tide permitted, Makarov put to sea with his entire force, but the Japanese were already gone. He thereupon steamed five miles out to sea to put the fleet through its first wartime exercises, indeed its first exercise in significantly longer than that. Results were not entirely encouraging. There were several instances of battleships side-swiping, but no serious damage was done.

170

After the fleet returned to harbor, Admiral Makarov at once set in motion a plan to station observers on the surrounding mountains with telephone links to the ships, so that in the future their fire might be directed from these positions ashore, and they might reply accurately to any future Japanese indirect bombardment. He also instituted a series of exercises designed to decrease the amount of time required for the squadron to leave harbor. He was given twelve days' grace to perfect this maneuver; for that period of time the Japanese did not appear in daylight, and even their night destroyer patrols became a good deal more circumspect. In any event the time for the fleet to gain the open sea was reduced to two hours and thirty minutes, almost exactly one-tenth of the pre-Makarov time.

On the night of March 22, a torpedo attack was made against the harbor mouth; no damage was sustained by either the Russians or their assailants. The following day, the First and Second Battle Squadrons approached at first light. *Fuji* and *Yashima* stood in to open an indirect fire upon the town and ships, as had been done on March 10. Makarov's spotters went into action, and fire was returned by the forts, and by *Tsarevitch* and *Retvizan* (since the two battleships were not yet seaworthy). Still, by 0700 a sizeable Russian force was outside the harbor. *Askold* led the five battleships *Petropavlovsk, Poltava, Sevastopol, Pobieda* and *Peresviet.* (The Warners state that Makarov's flag still flew in *Askold*, whereas Westwood maintains that he had already transferred to *Petropavlovsk*.)

The Russian battleships opened fire upon *Fuji* and *Yashima* at 0930. The two Japanese at once drew off toward their consorts; the Second Battle Squadron's armored cruisers commenced firing and stood in to cover the withdrawal. For his part, deprived of his two newest battleships and observing the poor standard of Russian gunnery, Makarov declined to be drawn beyond the range of his shore batteries. *Fuji* was slightly damaged, but she and *Yashima* withdrew successfully, and the Japanese retired unpursued. Firing ceased by 1100, and by 1500 all Russian ships were safely back in harbor.

The facility with which the enemy departed and re-entered the harbor was not lost on the Japanese command. In consequence, Admiral Togo was instructed to prepare a blocking operation, making use once again of redundant merchantmen. Four aging tubs were

171

procured and fitted out, in view of the failure of the previous expedition (though with logic which it is difficult to penetrate) the size of the scuttling charges was increased to the extent that the ships may properly be spoken of as blowing themselves up.

At 1900 on March 26, the four set out, accompanied by destroyers. They proceeded without incident; the sea was smooth, the night calm and misty, the moon mostly obscured by cloud. At 0220 March 27, the destroyers fell astern and the block ships began the run-in.

At 0245 the Russians opened fire upon the approaching merchantmen. What occurred subsequently it is difficult to say with precision, for the scanty records are in a state almost as chaotic as that which must have prevailed that night, as searchlights probed and crossed and muzzle flashes stabbed out into the dark. Hargreave asserts that one ship, *Yonemaru Maru*, blew herself up "in the fairway" at precisely the moment she was torpedoed, but that she proved only a partial obstruction. The other three, he states, were sunk by gunfire at a safe distance from the port. The Warners attribute the partial blocking of the fairway to *Fukui Maru*, concerning which vessel they describe the same sequence of events as Hargreave attributes to *Yonemaru Maru*. They add that *Chiyo Maru* blew herself up "in the channel . . . near Golden Hill". This ship was commanded by Ryokitsu Arima, the officer who had conceived the block ship operation. It was the Japanese tradition that the author of a plan ought not to shirk the execution of it, and Arima accordingly laid down his life. I would suggest that, under the circumstances, if *Chiyo Maru* had in fact been destroyed by gunfire at some distance from her objective, the Japanese would have been loath to admit it. Thus, I suspect, the relative success the Warners, attribute to the effort as regard the *Fukui Maru/Yonemaru Maru* discrepancy is the result of drawing upon Japanese documentation. In the absence of an authoritative reference it is irreconcilable. With the failure of the blocking attempt, a comparative quiet descended upon the Port Arthur seafront. It was broken, as it chanced, by simultaneous but entirely separate operations on the part of the Japanese and the Russians; the coincidental meshing of the two efforts was to have profound consequences.

172

On the evening of April 11, Admiral Makarov dispatched eight of his destroyers to reconnoiter the Elliot Islands, about 70 miles north of Port Arthur. It was Makarov's suspicion that the Japanese fleet might have established an advance base there. (They did indeed intend to establish such a base, but the project had not yet proceeded beyond the planning stage, so the Russian scouts found nothing.)

The Japanese fleet was at any rate, not in port anywhere that evening. The entire force, including the newly-purchased armored cruisers *Kasu*ga and *Nisshin* (making their operational debut), was underway to escort the recently-fitted-out minelayer *Koryo Maru* to Port Arthur, Subsequently, the intention was to display light forces and entice the Russians out, then either engage them in a fleet action or maneuver them onto the new minefield.

At 1740 April 11, the Japanese came to anchor beyond sight of Port Arthur, and *Koryo Maru* with her auxiliary launches set out to mine the entrance to the Russian base. (At 2250 Admiral Makarov was personally in visual contact with the mining force, from the bridge of the anchored *Petropavlovsk*. Believing that the activity might be that of his own destroyers--"They have lost their way," he said, "and dare not return in the darkness for fear of being taken for Japanese."--he took no action beyond noting the spot and ordering that it be checked for mines at daylight. In the excitement next day, the order was overlooked. Forty-eight mines were laid, and the Japanese withdrew without incident.

Other Japanese forces were prowling about near Port Arthur in the darkness, and one of the Russian destroyer patrol, *Strashni*, had lost her formation. Seeing hooded stern lights, she closed up and joined formation, only to find as dawn broke that she was in contact with the Japanese Second Destroyer Division. Recognition and the commencement of firing appear to have been mutual. *Strashni* fired one torpedo, which missed; the second was detonated in its tube by a direct shell hit, killing many men and devastating the midships section. Firing was deadly at ranges as low as 80 yards; *Strashni* scored a hit which wrecked the bridge of *Ikazuchi*, and peppered the funnels of another Japanese destroyer. The Russian was however overwhelmed with shellfire, and soon went down, blazing from end to end.

News of the engagement reached Port Arthur, and the cruisers at once prepared to go to the rescue. *Bayan* was first at the scene, too late to succor *Strashni*, or to do anything other than pluck the five survivors from the sea and hurry the Japanese destroyers on their way with gunfire. The idea of pursuit was quickly set aside, for the gunfire drew the cruisers *Chitose*, *Takasago*, *Kasagi* and *Yoshino*, with *Asama* and *Tokiwa* in close support. These engaged *Bayan* from 8000 yards; the Russian turned to retire upon Port Arthur. But as *Askold*, *Diana* and *Novik* could be seen coming to her assistance, and as the Japanese seemed not disposed to close the range, *Bayan* instead returned fire and continued upon a course parallel to that of the enemy. Her three consorts formed up astern, and a gunnery duel at medium range ensued, with no clear advantage to either side.

At length, about 0800 April 12, Admiral Makarov came up in *Petropavlovsk*, with *Poltava* in company. Observing the arrival of the two heavy units, the Japanese began slowly to open the range, hoping to lure the Russians away from the shore and onto Togo's battle fleet, which had been advised of the enemy's sortie and was approaching at full speed. Makarov was not disposed to follow, and rather than endeavoring to close the enemy, contented himself with keeping the Japanese within gun range. In due course *Sevastopol*, *Peresviet* and *Pobieda* came out and formed up astern of *Poltava*. Still, conscious that he faced only a small portion of the Japanese fleet, Makarov declined to exploit his overwhelming local superiority, lest he be led afield and return to find the enemy between himself and his base. The long-range firing continued unproductively.

The whereabouts of the main Japanese fleet was soon made manifest, for visual contact was gained with the main squadron under Admiral Togo at a range of about 26,000 yards. The Japanese cruisers made for the main force, while the Russians came about and returned toward Port Arthur, *Bayan* leading, deliberately crossing the spot at which *Strashni* had sunk, in the hope of rescuing further survivors (none were found). The Japanese did not open fire, but paralleled the Russian course at the limit of visibility.

At 0943 *Petropavlovsk* ran onto one of the newly-laid Japanese mines. The muffled thud of the mine's detonation was followed instantly by the thunderous detonation of the battleship's forward magazine. The forecastle decking was torn and blown upward around the forward turret, and Makarov was killed together with every man on the bridge save the Grand Duke Cyril, who later reported the Admiral's death. Observers aboard the other ships of the squadron saw the flagship enveloped in a dense cloud of brown smoke, her foremast canted crazily. A second explosion was heard as the boilers blew, adding a burst of white steam to the dirty brown pall. The smoke drifted momentarily clear to reveal *Petropavlovsk* beginning to stand on her nose, her screws still spinning in the air. Then the after magazine detonated, blowing a gust of flame from every port and bulkhead, and the ship was gone. She went down in less than a minute, taking 635 officers and men; 80 were saved.

Command devolved upon Rear-Admiral Prince Ukhtomski in *Peresviet.* He signaled the squadron to follow in single line ahead, and continued to make for Port Arthur. There was no certainty in the fleet as to the cause of the flagship destruction, and considerable ammunition was fired off at floating debris and driftwood, mistaken for periscopes.

At about 1000 an underwater blast rocked *Pobieda*; the slab-sided battleship began slowly to heel over. She was extensively damaged but was able to maintain her place in line and return to harbor. She also had struck a mine, but this was not known to the Russian gunners, who may fairly be said to have panicked at this second reverse. Undisciplined and totally directionless firing quickly became general throughout the squadron, continuing for some minutes before the officers were able to reassert control.

Admiral Togo observed all this through his binoculars, standing silent and motionless on the bridge of *Mikasa.* Whatever his personal desires and emotions, strategic necessity remained foremost, and he refrained from exploiting the Russian confusion. Around him, his seamen cheered and threw their caps into the air, but the Russians were permitted to return to port unmolested. By 1200 they had done so. Togo withdrew in his turn, leaving light forces on patrol.

Much has been made of the profound moral effect which the loss of Makarov, who had been known to them as "the little father", or more familiarly "beard face", took upon the Russian seamen. Indeed it does seem to have sapped the fighting spirit, though it must be said more among the officers than rank and file. Under another commander than the lackluster Vitgeft, who was appointed to succeed Makarov, the story might have been different. Still, it is necessary to reflect that Makarov's death was inherent in his tactics; the loss of *Petropavlovsk* was not mere mischance. The Japanese had laid the mines with the deliberate intention of bringing about just such an event. Had they not claimed the admiral on this occasion, they must have done so soon enough. Although as his fleet was still inadequately trained he maneuvered with considerable caution, even so his too literally Nelsonian dash took insufficient account of the underwater perils that had matured in the near-century since Trafalgar, and his death was all but foreordained.

Command of the squadron now devolved upon Viceroy Alexeiev, who was also responsible for the civil and overall military command of the threatened base. He was a bureaucrat, timid or possibly overtly treasonable; Admiral Stark had been his protégé, and his policy had the same "no-risk" basis as that officer's.

On April 15, the Japanese bombarded Port Arthur for two hours without encountering any opposition from the Russian ships. Two days later the new *Kasu*ga and *Nisshin* closed in to calibrate their guns against the Russian works. It was the sort of provocation which had customarily brought Makarov out at a run, but now the Japanese shooting was interrupted only by the counter-fire of shore batteries. The brief revival of the Russian squadron was ended.

It had been intended to send some qualified officer from the Baltic to replace Makarov, but growing Japanese success on land combined with their virtual command of the sea approaches made the risk of sending in a new admiral too great to be entertained. The senior officer present. Admiral Vitgeft, was appointed to the post. His remarks upon taking up the appointment were unfortunate; all would be expected to "aid me in word and deed," for "I am no commander of a fleet", as he blurted out to his nonplussed subordinates.

Vitgeft at once made clear his intention not to risk fleet units beyond the confines of the harbor. He intended instead to render Port Arthur as nearly as possible impregnable from the seaward side. It is only fair to say that in this he was virtually successful; on the other hand the Japanese had no intention of forcing an entry to the harbor by sea. Additional block ships were sunk in the fairway by the Russians to further impede access to the narrow channel without actually immobilizing the Russian ships within. Some of the ancient gunboats were moored as floating batteries in such positions that their guns bore on the most vulnerable points in the course of the approach; and *Askold* and *Bayan* were so positioned within the anchorage that their guns swept the fairway.

Togo, although not yet sure enough of Russian inactivity to sanction the troop landings the army wanted to make to seal off Port Arthur from the landward side, was nonetheless prepared, on April 22, to detach Admiral Kamimura with the Second Battle Squadron (the six armored cruisers, *Kasu*ga and *Nisshin* excluded) together with light units to mine the entrance to the harbor of Vladivostok. The four Russian cruisers stationed at that port were a constant threat to Japanese merchant traffic.

As it proved, the measure was late; for the Russians were already at sea. In dense fog on the morning of April 26, *Rurik*, *Rossia*, *Gromoboi* and *Bogatyr* intercepted two Japanese troop transports. *Kinshu Maru* and *Goya Maru*, carrying elements of the 57th Infantry Regiment, were halted and given one hour to surrender. The crews took to the boats; the soldiers lined the rails and opened a futile rifle fire against the Russian warships. Numbers of them also chose to commit suicide rather than face surrender. At the expiration of the hour, the Japanese soldiers showing no inclination to abandon ship, the two transports were torpedoed and sunk. The Warners state that not one Japanese soldier survived; Hargreaves gives figures of 100 killed and 250 taken prisoner.

In the meantime, also in fog, Kamimura's ships had duly laid 75 mines in the vicinity of Vladivostok and departed. As it proved, these did not inconvenience the Russian ships on their return, or indeed at any time. Pressure had mounted upon Admiral Togo to immobilize the Port Arthur Squadron so that troop landings could be made in the vicinity to establish a siege perimeter around the base. Accordingly, a third attempt to seal the anchorage with block ships was proposed and accepted. Twelve ships were chosen for the assault; two were disabled in heavy seas, but the remaining ten reached the fleet. On the evening of May 3, they began the run-in under heavy cloud with the sea rising.

In view of the increased difficulty of the approach, torpedo boats were stationed at difficult turns along the way, showing hooded lights to guide the block ships in. *Shiheta Maru*, commanding the flotilla, attempted to abort the mission on account of the weather, but only two ships saw and obeyed the recall signal.

As we have seen, this was the one occasion for which the otherwise lackadaisical Vitgeft had prepared and by the third time around, the affair must have seemed like an exercise to the Russian gunners. At 2400 Russian searchlights snapped open to reveal the seven ships as they bore down, and the guns, already trained in, at once began slamming shells down the fairway. The Japanese with their massive "scuttling" charges went up in pyrotechnic splendor, the last one sinking 1800 yards from the entrance to the harbor. Sixty-three Japanese survived out of a total of two hundred thirty four aboard the seven ships.

Surely the results of the foray were reported accurately to Admiral Togo, yet he, for whatever reason, pronounced the harbor sealed, and on May 5 the 2nd Army, commanded by General Oku, began landing at Pitzuwu, 40 miles northeast of Port Arthur. (As we know, the psychology of Admiral Vitgeft made the landings every bit as safe as if the entrances to the harbor had in fact been blocked: Admiral Togo could not have known this, and he must have felt considerable anxiety concerning the possibility of a descent upon the amphibious forces by the Russian ships. He did what he could to step up and intensify patrolling activity off the Russian harbor, but his blanket assurances to the army must have weighed heavily upon him until the troops were securely lodged ashore.)

On May 12, the Japanese *TB48* was lost on a mine near Dalny; enough mining had taken place so that it is not possible definitely to assign the source of this particular one. It may have been one of the number of mines that had parted their moorings and were drifting free, a menace to friend and foe alike.

Meantime, Japanese battleships paraded daily before Port Arthur, maintaining a distance of about 20,000 yards from the Russian works. Admiral Vitgeft refused to yield to the requests of his men to be allowed to engage the enemy units, and forbade any Russian vessel to venture more than 14,000 yards from the anchorage. Nevertheless, on the night of May 14 the minelayer *Amur*, ignoring this stricture, proceeded to sea and laid a string of mines along the usual Japanese maneuvering-ground.

In the pre-dawn darkness of May 15, as units of the blockading force proceeded toward their Elliot Island base, *Kasuga* rammed and sank *Yoshino* in error. The gallant old veteran was riven beyond saving, and went thus ignominiously to the bottom, carrying with her 518 officers and men. The ill-omened day had scarcely begun.

As daylight came, *Hatsuse*, *Shikishima* and *Yashima* approached Port Arthur and commenced the daily observation patrol. The captain of *Amur* had meanwhile confessed his foray to Admiral Vitgeft; we may surmise that his attention was not entirely upon the angry remonstrances of his commander, for from the bridge of the flagship he could see the enemy battleships steering straight into his newly laid field.

As she crossed before the harbor at 1030, *Hatsuse* was shaken by a powerful explosion right aft, which destroyed her propellers and flooded her steering compartment. The stricken vessel drifted to a halt. *Yashima* came up to pass a tow, but at 1035 she also ran onto a mine, and was "enveloped in smoke and flames". It was not as serious as it looked; *Yashima* ran out of the cloud of smoke, listing badly but still underway, albeit making a fraction of her normal speed. *Hatsuse* was dead in the water, but *Kasagi* made ready to pass a tow, while *Shikishima* endeavored simultaneously to mount guard over the damaged battleships while avoiding running upon a mine herself in the process.

Ashore, Russian jubilation was at first unbounded; yet the triumph turned bitter as it became evident that, even with major enemy units immobilized in full sight, Vitgeft had no intention of venturing forth to complete their destruction. It is from this event that the true collapse of Russian naval morale may be dated. The men lined the rails, watching the discomfited enemy struggling to save themselves, raging at their self-imposed impotence; and after this, they were simply never the same.

At 1233, *Hatsuse*, still wallowing powerless and uncontrollable, drifted onto a second mine, which detonated her magazines: she exploded with such violence that her masts and funnels toppled, and she sank by the stern in less than a minute. All who could be spared below were topside, since the ship was already in distress, so that about 300 men were saved, while 493 lost their lives. The Russian cheers rang hollow.

Towing 180 men astern in boats to simplify abandonment if necessary, *Yashima* plodded painfully along, seeking with agonizing slowness to escape Russian observation. Her captain was already assured that she would eventually be lost, but felt that it might be possible to conceal the loss from the enemy if she could be kept afloat and gotten out of sight by dark. Admiral Vitgeft at last dispatched *Novik* with a flotilla of sixteen destroyers; their orders were "to worry not to attack", and they contented themselves with firing on the Japanese before retiring in the face of aggressive enemy torpedo craft.

By 1600 *Yashima* was beyond Russian vision (and indeed her loss was kept secret until after the end of the war), but within an hour she was dead in the water and it was clear her end could not be long delayed. Confidential papers and the portrait of the Emperor were got away, and the crew, attired in fresh uniforms, lined the rail to sing "Kimigayo" before receiving the order to abandon ship. They did so without loss, and sometime between 1800 and 2000 *Yashima* capsized and sank.

Still on this dire 15th of May, the Russian protected cruiser *Bogatyr* ran aground off Vladivostok, damaging herself so severely that she was out of the war. It was a meager compensation for Japanese losses, and the Japanese were not even aware of it.

Two days later, the fine modern destroyer *Akatsuki* struck a mine while maneuvering 16,000 yards off Port Arthur. She went to the bottom with 45 officers and men. (With the intensive Japanese mining of the approaches to Port Arthur, the number of free-drifting mines increased, for it was inevitable that a portion of those laid would separate their moorings It cannot be definitely asserted that such a mine destroyed *Akatsuki*, but in view of Russian inactivity it is probable that this was the case.)

On May 25, the Japanese Army stormed and took Nanshan Hill, a key position in the outer ring of Port Arthur's landward defenses. The Russian base was now cut off, and from this date we may properly speak of the siege of Port Arthur.

KIMIGAYO

A word about *Kimigayo*, sung by the crew aboard *Yashima* before they abandoned the vessel, not because it is important at this point, but because it is interesting and there is no other place where it will fit well. This was the de facto national anthem of the Empire of Japan, returned officially to this status in 1999. It is among the shortest national anthems, being only 11 measures and 34 characters in length. The lyric was written in the Heian period (794–1185), the music not until 1888. The title is usually translated *His Imperial Majesty's Reign* and it is a highly nationalistic piece offering insight into the Japanese mindset during its period of imperial expansion. This fact and its size allow me to insert it in its entirety here:

> May your reign
> Continue for a thousand, eight thousand generations,
> Until the pebbles
> Grow into boulders
> Lush with moss

Translated by Christopher Hood in 2001

The emperor and his empire will last for eight thousand generations or until pebbles (the islands of Japan?), grow into moss-covered boulders, (like other world powers particularly China?). It is worth remembering that Japan is the outside world's name for their country, derived, as is the rising sun symbolism and flag, from Chinese perceptions of this island nation to the east. Japan is where the sun rises for China, thus it became the land of the rising sun for China and ultimately for the rest of the world. The Japanese refer to their country as Nippon of course, meaning the origin of the sun, of power and perhaps the entire universe. Little wonder that this philosophy drove Japan's imperial ambitions and destiny in this era of imperialism and manifest destiny.

Compare this to *Rule Britannia,* discussed on page 12 and 13.

SIEGE OF PORT ARTHUR,
MAY 25, 1904 - JANUARY 2, 1905

As soon as the siege proper commenced, Viceroy Alexeiev, and above him the person of the Czar, began to pressure Admiral Vitgeft to break out to Vladivostok, join the cruisers there, and operate the fleet against Japanese communications. This demand the dilatory commander put off as long as possible (although in fairness it should be remarked that the majority of ships' captains were also in favor of remaining to add their guns and men to the defense of the base. We know this because Vitgeft had characteristically canvassed them for their opinions.) At length, assured that repairs to his damaged units would complete at that time (thus depriving him of his chief excuse for inaction) Vitgeft ordered on Jun 15 that the squadron should be ready to sail on June 20.

On June 12, *Gromoboi*, *Rossia* and *Rurik* sortied from Vladivostok to prey on Japanese shipping. On June 15, two merchantmen carrying eighteen 11-inch Krupp siege howitzers, as well as locomotives and rolling stock, were intercepted and sent to the bottom. The same day *Gromoboi* shelled and sank the troop transport *Hitachi Maru*; 1300 of the 2000 soldiers aboard were killed or drowned; the remainder made shore in the ship's boats. Upon receiving news of these reverses, Admiral Togo dispatched Admiral Kamimura with *Asama*, *Tokiwa*, *Iwate, Yakumo*, *Adzuma* and escorting torpedo craft to track down the marauders. In this they were unsuccessful, but in consequence of the operation, they were still absent when the Russians came out, and it was a grievously truncated Japanese line which the enemy faced.

The sortie, planned for June 20, was postponed at the last minute when it was learned that a newspaper report of the fleet's impending departure had been published. It was not until 0400 June 25 that Vitgeft ordered his ships to raise steam and prepare for sea. (Doubtless he hoped by keeping the news even from his officers to preserve secrecy and so escape without encountering the enemy fleet. In this he was disappointed by the slowness of his operations, for since Makarov's time the hazards of mines and block ships had multiplied,

182

and the habit of quickly leaving the harbor had been lost.) *Vsadnik* and *Gaidamak* swept a channel for the departure; in the first hour eleven mines were destroyed, superficially damaging both sweepers by their explosions. Still, it was 1200 before the major warships got underway.

As the object of Russian movements became obvious, some of the patrolling Japanese destroyers made for the Elliot Islands to alert the main fleet, while the others closed in and opened fire upon *Vsadnik* and *Gaidamak*. *Novik* at once advanced in support, driving back the Japanese, and the sweeping continued without hindrance. At 1630 the fleet at length proceeded to sea in line ahead in the order *Tsarevitch* (flying Vitgeft's flag), *Retvizan*, *Pobieda*, *Peresviet*, *Sevastopol*, *Poltava*, *Bayan*, *Pallada*, *Diana* and *Askold*. *Novik* and seven destroyers proceeded to starboard of the line.

At first all seemed to go well; but at a distance of 40,000 yards from the base, the Japanese fleet was sighted at about 1800. They were in line ahead in the order *Mikasa*, *Asahi*, *Fugi*, *Shikishima*, *Nisshin*, *Kasu*ga, *Idzumi*, *Akashi*, *Akitsushima* and *Suma*. Thirty destroyers and torpedo boats were in company, and lurking in the wings were a further eleven obsolete warships, including the antiques of the *Hashidate* class and the utterly obsolete ex-Chinese battleship *Chin Yen*.

At first the Russians held on steadily enough, and by 1845 the fleets were steaming in parallel, 16,000 yards apart. *Mikasa* altered course to close, but at 1900 *Tsarevitch* came right about and led the fleet back toward Port Arthur.

Togo came about as well, and for a short time looked like pursuing; but prudence governed, and he turned away again, unleashing only his torpedo craft to harry the retiring foe. Meantime, the Russians huddled back toward Port Arthur in a dispirited mob. *Novik* and the destroyers were released to make for home at their best speed, and the cruisers came up along the starboard sides of the slower battleships and gradually took the lead. The Japanese torpedo attacks were unsuccessful, but the additional haste and confusion which they inspired among the Russians meant that the swept channel could not properly be located in the twilight, and *Sevastopol* struck a mine. The stricken battleship put into White Wolf Bay and hung her torpedo nets; the rest of the squadron anchored in the roads of Port Arthur, unwilling to risk the entrance channel in the darkness. Torpedo nets were hung, searchlights switched on, and gun crews held at their stations.

Beginning at 2045 and continuing until dawn of June 24 Japanese torpedo craft attacked at will. The blaze of illumination and the heat of the Russian gunfire kept the enemy at a respectful distance, none coming closer than 6000 yards, and in consequence none of the Russian ships was struck. Neither did the Japanese sustain any damage worthy of mention. The following day the Russians re-entered the port without any further attempts by the enemy to inconvenience them. Still, despite the abortive nature of the encounter, the Russians felt thoroughly defeated, while for the elated Japanese it was just the opposite.

Indeed the feelings were justified; Vitgeft, in his fixed insistence upon a surprise escape and his unwillingness to contemplate action, had cast aside a unique chance to engage Togo with little more than half his force in company. The Japanese, on the other hand, were quite well aware that they had met a major Russian attempt at a breakout, and had turned it back with scarcely a shot fired. Russian morale plummeted still further, (in truth there had been an ugly panic verging on insubordination aboard *Sevastopol* when she struck the mine, and only the iron nerve of Captain Essen had preserved order).

While it is quite true that henceforth many of the men and a portion of the secondary and tertiary armaments of the ships fought ashore, the fleet as such was allowed to lapse into a gloomy torpor. On June 27 *Bayan* was gravely damaged by striking a mine; this set-back curtailed even such minimal patrolling by major units as had gone on. There was still a continual prowling and counter-prowling of smaller craft; in particular, as the month of July wore on, a nest of destroyers at Ta Ho Bay was making itself a nuisance to the Japanese. On the night of July 24, boats from *Mikasa* and *Fuji* were fitted out with torpedoes and sent to penetrate the bay. The three Russians present, *Leitenant Burakov*, *Grozovoy* and *Boevoi*, did not detect the diminutive enemy launches. The Japanese made their way along the shore until they could fire on the Russians from the landward side, which they then did to considerable effect. *Leitenant Burakov* was blown in half and sunk outright; a torpedo detonated in the forward stokehold of *Boevoi*,

184

inflicting serious damage. The unhurt *Grozovoi* failed to prevent the withdrawal of the audacious Japanese.

In the meantime, things ashore had gone ill for the Russians, and Japanese land artillery had begun to drop shells into the harbor. The ships fired back, but it was a losing proposition. *Retvizan* took seven hits, which caused the ship to take in 400 tons of water; *Peresviet* was also hit several times. It was obvious that the situation was untenable, and on August 7, the order was received, "Put out with full strength for Vladivostok," over the signature of Czar Nicholas II. It was proposed by subordinates that the less battle worthy *Poltava* and *Sevastopol* be detached to bombard the Japanese supply depot at Dalny. They were units of secondary importance, being old and slow, and there was the chance that they might draw the major Japanese effort and facilitate the escape of the rest. The plodding Vitgeft, profoundly depressed and plagued with premonitions of death, rejected this scheme, pointing out that the Czar's order specified the "full strength" of the squadron.

Admiral Togo anticipated that the Port Arthur squadron would attempt to reach Vladivostok, and also that the Vladivostok cruisers would sortie in support. Therefore, and such was the slight esteem in which he held his opposition, he deliberately accepted the same odds which circumstance had forced upon him on June 25; that is, he dispatched Kamimura with his-four armored cruisers to intercept the Vladivostok force, choosing to face the main enemy force with his four battleships, along with *Kasu*ga and *Nisshin*.

THE BATTLE OF THE YELLOW SEA,
AUGUST 19, 1904

Within seventy-two hours the Russians were prepared to sortie, and at 0421 August 10, they set out, preceded by minesweepers. Left ashore to bolster the landward defenses were ten 6-inch and twelve 12-pounder guns. Also, one of *Sevastopol*'s 12-inch guns was damaged, and had been replaced with a wooden replica. The flagship *Tsarevitch* led off, followed by *Retvizan, Pobieda, Peresvet, Sevastopol, Poltava, Pallada, Diana* and *Askold* (*Bayan* being left behind because of mine damage). *Novik* and eight destroyers were in attendance, with eight more destroyers and minesweepers which were not intended to accompany the sortie.

The day dawned fine and calm, with a light breeze and low-lying mist. The Russians were clear of Port Arthur by 0800. *Ryeshitelni* was detached to Chefoo with dispatches, the minesweepers returned to the port, and the rest moved off toward Vladivostok. The Japanese were meanwhile idling off Round Island, *Mikasa, Asahi, Fuji, Shikishima, Kasuga* and *Nisshin* (*Asama* being absent coaling). In attendance upon the six heavy ships were seventeen destroyers and twenty-nine torpedo boats.

Admiral Dewa's flag flew in *Yakumo* as his detachment of cruisers, scouting the entrance, sighted the enemy as they came out and shadowed them, reporting their movements to Admiral Togo at Round Island by wireless (the first combat use). One 12-inch shell from *Poltava* struck *Yakumo*, but shadowing continued until the main units came up; by 1100 the fleets were in visual contact at 24,000 yards. The Japanese approached the Russians, then at 1133 turned 90° to starboard to cross ahead, then in quick succession formed line abreast moving ahead of the Russians, and turned once more to port to form line ahead in reverse order. *Tsarevitch* and *Retvizan* opened fire briefly at a range of 8000 yards. The Japanese did not reply, and the Russians presently swung sharply to starboard, fearing that the Japanese had dropped mines in their wake. Thus, the range opened once more, and firing ceased. A period of confusing maneuvering followed, with the Japanese attempting to force the enemy back into Port Arthur. Vitgeft,

186

maneuvering to evade any floating mines (though in fact the Japanese dropped none), and obstinately held out for Vladivostok.

At 1230, as the Japanese crossed once more ahead of the Russians, Vitgeft cut sharply to starboard and passed astern of the enemy formation, firing upon *Asani* and *Mikasa* at 7000 yards in passing. Togo turned together to retake his place at the head of the line, and by 1500 the fleets were passing on opposite courses, with firing general at the extreme range of 14,000 yards. At 1330 Togo turned in succession to cut off the Russians once more while still maintaining *Mikasa*'s place at the head of the line. This over-lengthy maneuver gave Vitgeft an opening, and he darted through and made off at full speed toward the Straits of Korea. *Askold* had sustained shell damage, but his ships were otherwise unhit and he had a 10,000-yard lead on his pursuers.

The Japanese pursued on a parallel course, rather than steering in the enemy's wake; the Russian battleships were equipped to drop mines astern, but there is no evidence that any did so on this occasion. The Japanese gunners attempted to concentrate upon *Tsarevitch* at the range of 10,000 yards, but firing trailed off as its ineffectiveness was noted. The ships settled into the chase, Togo relying upon his extra knot of speed to overtake the Russians before nightfall. The stokers were the heroes of the afternoon, laboring below decks in temperatures that reached 130°F. Topside it was a magnificent spectacle, the slipstream snatching the smoke astern directly it left the funnels, the deck throbbing beneath the feet, the ships pounding along beneath the clear summer sky.

The pursuit went well enough, and by 1525 the Japanese were far enough ahead that Togo once more altered course toward the Russians. At 1600 firing resumed at 10,000 yards and continued as the range fell. Three 12-inch shells struck *Mikasa*, one at the water, one bursting against the after funnel, and the third blowing off the muzzle of one of the aft 12-inch guns, disabling the turret; 31 men were killed and 94 wounded. A shell also struck the bridge of *Nisshin*, killing 16 and wounding 15 men. The Russians did not escape unscathed, and by 1750 there were four Russian 12-inch guns out of action compared to five Japanese, firing mishaps as well as shell hits being responsible for some of these. *Peresviet*'s topmasts were down, meaning that Admiral

Prince Ukhtomski, the second in command, was unable to signal effectively. It had been an even match, and the Russians looked to have a fair chance of escaping.

Thus far the performance of the major-caliber ammunition had been disappointing, the "armor piercing" shells of both sides having the distinct tendency to burst on impact. Now this defect was to prove critical to the outcome of the battle, for a 12-inch shell struck the unarmored bridge of *Tsarevitch*. Rather than slamming through to burst below, the projectile exploded upon striking, killing Admiral Vitgeft (only his left leg could be identified) and all with him. A second 12-inch followed upon the path of the first, detonating against the armored wall of the conning tower. It failed to penetrate, but the force and reverberation of the explosion concussed those within insensible.

Among them, the helmsman fell onto the wheel, jamming it hard over to port. *Tsarevitch* swerved away from the enemy with no signal of intent, circling completely around to run down upon her comrades. *Retvizan* tried to conform to this movement, while *Peresviet* and *Sevastopol* were compelled to evade violently as the flagship broke through the line between them. The Russian formation broke down completely. At length, signaling with difficulty, Ukhtomski formed the ships into line ahead once more, swung them about to starboard and headed them back toward Port Arthur. (Surely the journey back was no less fraught with peril and uncertainty than that ahead to Vladivostok: Ukhtomski cannot have been ignorant of his sovereign's explicit command: the wrong-headed decision taken is so glaringly obvious that one must wonder about the persistent Russian allegations of deliberate treachery in high places. Such were the Byzantine intrigues of Tsarist Russia that this possibility was and still is, given very serious consideration.)

Tsarevitch continued to circle astern of the retreating squadron; only *Retvizan* remained in company. She had conformed to the flagship as well as she might throughout, and now stood and fought a rear-guard action of considerable brilliance while the flagship was brought under control and slipped off into the darkness toward Vladivostok. The Japanese closed to within 4000 yards, all guns trained upon *Retvizan*. The lone Russian, at last out of contact with *Tsarevitch*, followed in the main squadron's track back toward Port Arthur, erupting with the smoke and flame of shell bursts, and returning a vigorous fire against the pursuing Japanese.

Still, it would have gone badly enough with her, outnumbered as she was, had not Russian destroyers been sighted coming up in the twilight. The Japanese torpedo craft were miles astern, for they could not keep up with the heavy ships through the long day's chase, so rather than risk his heavy units to the Russian torpedoes, Togo broke off, and *Retvizan* pursued her course homeward.

In fact the majority of the fleet: *Retvizan*, *Peresvet*, *Pobieda*, *Poltava*, *Sevastopol*, *Pallada* and three destroyers made their way back into the doomed base. The remainder, driven by chance or the hope of obedience to the Tsar's order, broke away and escaped. The next day the destroyer *Burni* was intercepted by Japanese destroyers and driven ashore near Weiheiwei, becoming a total loss.

Askold and *Grozovoi* made the neutral port of Shanghai, where they were interned for the duration of hostilities in accordance with international law. *Diana* similarly interned at Saigon. *Tsarevitch*, *Novik* and three destroyers reached neutral Kiauhhiau; the battleship and the destroyers were interned, but the little-damaged *Novik* fueled and departed within twenty-four hours, as was her right, and made for Vladivostok. She passed between the islands of Hokkaido and Sakhalin undetected but on August 18, *Tsushima* intercepted her. The two shot it out to an even draw, breaking off at nightfall with both antagonists damaged. *Tsushima* kept contact, and the following day *Chitose* came up as well, allowing her damaged compatriot to withdraw. The outgunned *Novik* fought bravely, but was at length driven aground badly battered near Korsakowa Harbor. Her crew set scuttling charges and escaped ashore; it was not until July 21 1905 that the Japanese reclaimed the riven hulk. (*Novik* was remade into a dispatch vessel, with one funnel removed, as well as all but two of her 4.7-inch guns; she was reboilered, renamed *Suzuya*, and incorporated into the Imperial Japanese Navy, in which she served from 1906 to 1913.)

Meantime, on August 12, *Ryeshitelni* was located at Chefoo by Japanese destroyers. The Japanese took her after hand-to-hand fighting following a dawn boarding assault from small boats. She was incorporated into the Imperial Japanese Navy, renamed *Akatsuki*, and served until 1918.

THE BATTLE OF ULSAN, AUGUST 14, 1904

Concurrently, the Vladivostok Squadron had sortied in support of the break-out by the Port Arthur squadron, just as Togo had anticipated. It was not until August 15, that Admiral von Essen (not to be confused with Captain Essen of *Sevastopol*) was prepared to sail. Still in ignorance of the defeat and death of Vitgeft, he set out with his flag in *Rossia*, with *Gromoboi* and *Rurik* in company.

Early in the morning of the next day, the three were intercepted by Admiral Kamimura's squadron, *Idzumo*, *Iwate*, *Adzuma* and *Tokiwa*, off the Korean port of Ulsan. The Russians at once turned away and fled back to the north, with the Japanese in full chase. Firing began at once; the sun was behind the Japanese line, its glare blinding the Russian gunners. *Rurik*, oldest and slowest of the three, fell behind, and accurate Japanese shooting quickly began to tell against her. Presently her steering was shot out, and she began to circle helplessly, exposed to the fire of the four oncoming enemy cruisers. *Rossia* and *Gromoboi* doubled back to her aid, and the hottest fighting of the morning ensued. An 8-inch shell struck *Iwate*, detonating ready ammunition and destroying three 6-inch guns; 40 men died and 37 were injured by the explosion. *Adzuma* took about ten hits and *Idzumo* probably twice that; *Rossia* also had three guns knocked out, and *Gromoboi* was afire. It became obvious that it would be impossible to regain control of *Rurik*; *Rossia* made to her the impossible "Steer for Vladivostok" and the two newer vessels left her to her fate. As his ammunition was nearly exhausted, Kamimura allowed the two fleeing ships to escape, concentrating upon the destruction of the cripple. (Sometimes it is asserted that false reports of ammunition depletion persuaded Kamimura to follow this course of action in error.) *Rurik* had two guns still in action, and stood up gallantly to the overwhelming force of her opponents, but of course her fate was a forgone conclusion. At 0945 her last gun was silenced; the crew fired her last torpedo (a miss), opened her valves and abandoned ship. *Rurik* went to the bottom with 170 dead; the Japanese took 625 prisoners.

Rossia and *Gromoboi*, much damaged and with heavy casualties, returned to Vladivostok on August 16. Almost immediately, *Gromoboi* stranded upon a reef near the harbor, damaging herself so severely that she was out of the war. *Rossia* alone was regarded as too weak to sortie, so the influence of the Vladivostok Squadron upon the war was at an end.

THE DESTRUCTION OF THE PORT ARTHUR
SQUADRON
AUGUST – JANUARY, 1905

Back in Port Arthur, the returnees from Vitgeft's forlorn sortie were no longer even a fleet in being. The men fought ashore, and the warships were systematically stripped of their guns to reinforce the landward positions, 284 guns of all calibers being unshipped and emplaced ashore. Some patrolling continued, *Gremyashchi* being lost on a mine August 18, and *Vuinoslivi* similarly lost on the 24th.

Ashore, a relief attempt from the north had gone badly, and by September 3, the Russians were retreating back into Manchuria. About this time an order, smuggled into the port aboard a Chinese sailing vessel, instructed Vitgeft's successor, Admiral Viren, to break out with the remainder of his ships to Vladivostok. Viren refused to obey, citing the necessity of naval guns to the maintenance of the defense ashore, and the likelihood that all his ships would be sunk, to no purpose if he ventured forth. This was the end of the subject, and at any rate by this time the ships were little more than laid-up hulks, awaiting their fate, slovenly in appearance and all but abandoned. *Sevastopol* alone of the big ships was kept in fighting trim, more through the zeal and fortitude of Captain Essen than through any design of the higher command. It is interesting to consider what the outcome might have been if Captain Essen's attitude had been more common in the Russian fleets.

The noose ashore continued to tighten; at sea, *Stroini* was sunk on November 13th, and on the 16th, *Rastoropni* went to the bottom. After fanatical assaults and counter-attacks ashore, Hill 203, with its commanding view of the anchorage, fell to the Japanese on December 5. Spotters were installed to command the fire of the 11-inch siege pieces ashore, and they began at once to "pick off" the immobile warships. On December 5, *Poltava,* one of the immobile warships, was hit six times; one of the shells ignited her magazines. The fires raged until she sank upright, when they were extinguished by the inrushing water. On December 6, *Retvizan* was the target, taking thirteen 11-inch shells before she sank. Her anchor had been embedded in a roadway ashore, and she dragged it after her as she sank, ripping up the paved road. On the 7th *Peresviet* and *Pobieda* both went down, taking twenty-three and twenty-one hits respectively. That evening, *Sevastopol* got underway and left the harbor, taking up a position in the shallows where she would be concealed from the spotters ashore, and protecting herself with torpedo nets. Such destroyers as were still operational accompanied her to provide protection from Japanese torpedo craft. On the day of December 8, *Bayan* and *Pallada* went to the bottom, so that the 11-inch guns had run out of worthwhile naval targets; their shells continued to enact occasional toll, one suspects more by chance than design. *Vsadnik* went down on December 15th, *Amur* on the 18th, and *Bobr* on the 26th.

Meantime Japanese light craft had at once begun intensive night torpedo attacks against *Sevastopol*. In all, thirty-five Japanese torpedo boats took part in five assaults against the battleship. The most costly was that of the night of December 14-15, *TB53* was lost on a mine; during the retirement following the attack, *TB42* was intercepted and sunk by the destroyer *Serditi*. A total of fourteen boats were damaged in the five attacks, but *Sevastopol* was hit by two of the 120 torpedoes fired at her, and the damage disabled her as a fighting unit. As the siege drew to its conclusion, on January 2, 1905 *Sevastopol* was towed to sea with a skeleton crew of 100 men, her valves were opened in deep water, and she heeled over to starboard and in ten minutes went to the bottom in thirty fathoms

The remaining destroyers were ordered to break for freedom; *Razyashchi*, *Storozhevoi* and *Bditelni* were lost, while the others interned in various neutral ports. Scuttled at Port Arthur were *Gjlgit*, *Razboinik*, *Otvajni* and *Gaidamak*.

The following ships were eventually raised and refitted by the Japanese, being incorporated into the Imperial Japanese Navy under the names given:

Betvizan/Hizen;	*Peresviet/Sagami*;
Pobieda/Suwo;	*Boltava/Tango*;
*Bayan/*Asc;	*Yariag/Soya*;
Pallada/Tsugaru;	*Xsadnih/Makigumo*;
Gaidamah/Shikinami.	

With Russian naval power in the East extinguished, Togo's ships were at length free to refit and prepare to meet the Second Pacific Squadron, which they knew to be on the way from the Baltic. The epic voyage of this ill-assorted force has been well documented, most notably by Richard Hough in his *The Fleet That Had To Die*. As the episodes of this long journey are not directly relevant to the topic of this book, I shall pass over them and begin my account of the climactic battle with the initial contact between Russian and Japanese warships.

BATTLE OF TSUSHIMA, MAY 27 - 28, 1905

The problem of intercepting the approaching squadron was simplified by the knowledge that they must come from the south, and that to reach Vladivostok, their only possible destination, they must pass around the home islands. Togo suspected that they would come by the shortest route, as indeed they did; therefore he took the chance of basing his fleet near Masan in Korea, closer to the southern straits. Cruisers and patrol craft operated from Tsushima, while fifty miles to the south a screen of Armed Merchant Cruisers (in fact fast liners) formed a distant-warning screen.

It was at 0245, May 27 that *Shinano Maru*, a ship of this screen, sighted the lights of the hospital ship *Orel* (not the battleship of this same name). *Orel* was correctly illuminated as a hospital ship, and was correctly identified as such, but her detection led to the discovery of the Russian fleet. Having made sure of the identity of her contact, *Shinano Maru* made by wireless "Enemy fleet in sight in square 203. Is apparently making for eastern channel." This was Togo's first indication that his gamble had paid off.

The main body of the approaching Russian squadron was deployed into single line ahead in three divisions: *Suvarov* (flag), *Alexander*, *Borodino*, *Orel* (the battleship), *Osliabia*, *Sissoi Veliki*, *Navarin*, *Nakhimov*; *Nikolai*, *Apraksin*, *Seniavin*, and *Ushakov*. The cruisers and destroyers were in column astern and to starboard of the battle line, conveying the auxiliaries and colliers that had accompanied the fleet on its epic voyage.

The morning dawned over thinning mist; the Russians proceeded at ten knots. About 0800 the Japanese cruiser *Idzumi* made visual contact with the Russian fleet and shadowed them from 9000 yards until, after an hour, *Suvarov* trained guns upon her, whereupon she made off into the mist. At 1000 Admiral Kataoka, with the ancient *Chin Yen*; and three cruisers, appeared to port of the squadron, but conscious of their inferiority they soon vanished again. Weather was worsening, with a rolling sea and the mist thickening once more. At 1200 Admiral Dewa's four cruisers made visual contact from the port beam of the Russian fleet and shadowed persistently. *Orel* at one point fired upon them at 9000 yards, and some of the other ships opened fire

195

as well; but *Suvarov* made "Ammunition not to be wasted", and at any rate the Japanese opened the range at once, so firing trailed off.

At 1300, anticipating the appearance of the Japanese battle fleet from the north, Rodjestvenski decided to form his First Division into line abreast to starboard of his line-ahead Second and Third Divisions, so that he would be situated to form line ahead in either direction as convenient to engage the enemy on his appearance. It is symptomatic of the poor discipline of the squadron that it proved impossible to execute the maneuver. *Alexander* misread the signal and followed in the wake of *Suvarov*, and *Borodino* and *Orel*, believing that they had misread the signal, followed *Alexander*. Thus rather than a formation of *Osliabia*, *Orel*, *Borodino*, *Alexander*, *Suvarov* abreast with the remainder of the fleet in line ahead behind *Osliabia*, the Russian formation consisted of two lines ahead, with the First Division to starboard and slightly ahead of the Second and Third. While Rodjestvenski attempted to sort out the muddle, visual contact was made with the Japanese battle fleet, coming from the north at a distance of 14,000 yards.

The Russians had by accident assumed a formation much like that which Togo anticipated to find, for the Russians had cruised in an order like the present one, and so his spies had reported them. He desired to attack the formation from the port side, because that side was weaker, Also being to windward it would give him the advantage of spray and smoke; with the rising sea, shell-holes on the windward side would damage the enemy more seriously than lee-side holes would damage the Japanese. Due to the cumulative errors of reckoning on the part of the scouts and of the fleet navigator, however, he overshot his mark, and visual contact found him already to starboard of the Russian line.

The Japanese were in line ahead in two divisions, *Mikasa* (flag), *Shikishima Fuji*, *Asahi*, *Kasu*ga, *Nisshin*; and *Idzumo*, *Adzuma*, *Tokiwa*, *Yakumo*, *Asama* and *Iwate*. The Japanese crossed before the oncoming Russians from starboard to port while still out of range at 15,000 yards. *Mikasa* swung to port, the others turning in succession behind her, and the line bore down on the port bows of the Russians on an opposite course, the time being about 1430. The Russians were attempting to sort out the confusion and return to single line ahead with the First Division leading; their success was not outstanding, and all the while the enemy bore down.

At length, with the range down to 8000 yards, Togo turned his ships in succession, reversing his direction but retaining the order of the line, so that each ship turned as if around a standing buoy. "How rash!" exclaimed a Russian officer, and it was all of that. The windy corner provided the Russian gunners with a stationary aiming-point, while at the same time the leading Japanese ships masked the fire of those following.

In the meantime, the Russian First Division, trying to squeeze back into the head of the line, had thrown the Second and Third Divisions into some disorder. *Osliabia* in particular was obliged to go dead slow, and was even so forced out of the line slightly toward the enemy. The gunners, however, were not slow to grasp their opportunity, and general firing began as *Mikasa* steadied onto her new course and *Shikishima* made the turn. Fifteen shells crashed into *Mikasa*, with a further fifteen hits in the first thirty minutes of action. As the greater portion by far of the Japanese damage was sustained at the commencement of action and within the succeeding half-hour (sixteen minutes of which were taken up with the turn), so we may as well summarize it here. *Mikasa*'s wireless aerial was shot away; *Shikishima* suffered ten hits, losing 13 men killed and 24 wounded; *Asahi* took nine hits, including a direct 6-inch shell hit on a 12-pounder position, which was demolished, 8 killed, 25 wounded; *Nisshin* suffered eleven hits, which put three of her four 8-inch guns out of action, killing 6 and wounding 14; *Idzumo* took nine hits; *Adzuma*, sixteen hits with 11 dead and 29 wounded; *Tokiwa*, nine hits, 1 killed, 14 wounded; *Iwate,* sixteen hits.

After the four battleships had made the turn, the Japanese replied to the Russian fire. *Suvarov*, with Rodjestvenski's flag, and the slab-sided *Osliabia*, thrust out toward the enemy and flying the flag of Rear-Admiral Folkersam, drew the brunt of the Japanese fire. Both suffered heavily, with the excessive superstructure of *Osliabia* quickly reduced to rubble and her forward 10-inch turret knocked out by medium-caliber shell hits. While *Osliabia* flew Folkersam's flag, the admiral had apparently died three days earlier of cancer, not battle, but his death was kept from the crew in an attempt to maintain morale.

The first ship of either side to leave the line, however, was the Japanese cruiser *Asama*; a 12-inch shell from *Nikolai* crashed in aft, demolishing the steering compartment. Two other 12-inch scored as well, out of a total of eleven hits, and the cruiser hauled off with 5 dead and 13 wounded. She was back in action within two hours. Meanwhile, *Osliabia* had taken three 12-inch shell hits forward, near the water and so close together that they formed one ragged hole. The leading edge of her armor belt was torn free of the hull framing, forming a veritable gate through which water poured into the hull. An attempt at counter-flooding merely brought her down by the bows that much faster, until water began rushing in through the side gun ports and she took an increasing list to port. Men swarmed up from below and over her side and keel as she rolled over and at 1530 went to the bottom, carrying about 600 men with her.

The line meantime attempted to hold on for Vladivostok, but the intensity of Japanese fire, and the tendency of the faster enemy to cross ahead, forced it ceaselessly away to starboard. At 1450 the battered *Suvarov* sheared away to starboard out of control, her forward funnel and both masts down, flames glowing through her ports and belching up from scuttles and hatches. A blast of shell splinters at 1500 swept the bridge, seriously wounding Rodjestvensky in the head, back and legs, and no sooner had he been hustled into the conning tower than a direct hit shook that structure, knocking him unconscious. *Suvarov* circled once around, and then staggered off to the northeast, imperfectly controllable. She was lost to the sight of the enemy, for the worsening weather, combined with the blowing banks of coal and gun-smoke rendered visibility slight and spotty. (The semi-conscious Rodjestvenski was at length taken off the wreck of the *Suvarov* along with members of his staff by the destroyer *Buini*, which in the absence of an intact boat had to close right up to the side of the battleship, a touchy maneuver. Later transferring to *Byedovi* as *Buini* lacked fuel, the admiral was taken prisoner the following day. *Suvarov* drew some fire from passing Japanese during the afternoon, withstood a torpedo attack, but succumbed to a second attack and went to the bottom around 1900, still firing her last 12-pounder. Other than the admiral and his staff, there were no survivors.)

Alexander, now leading the line and already afire, attempted a sharp turn to port to cut behind the Japanese line and make away toward Vladivostok. Togo's First Division duplicated her maneuver to hold her in thrall, while Kamimura with the Second Division continued on, passing the Russians at 5000 yards, then turned and followed the First Division. It was about 1500, and the Russians, continuously attempting to resume course for Vladivostok were foiled by the superior speed of the Japanese as Togo repeatedly headed them off. In the course of the battle, the Russians were run through two complete circles to starboard before settling temporarily on a course away from Vladivostok. As they passed clear of the worst of the smoke, Togo picked them up again, and the two fleets on parallel courses at 7000 yards shot it out for about a half hour from 1650. Togo then disengaged deliberately, ordering a torpedo attack against the Russian battleships, and firing ceased at 1635.

Learning that his destroyers were not in position to attack the enemy, Togo at 1655 hurried south once more toward flashes of gunfire on the horizon. These proved to be not the enemy battle line, but desultory fighting between his own and the enemy's light cruisers, who had been dueling inconclusively since about 1200. Kamimura was detached to aid the light cruisers, while Togo with the First Division stood back up in search of the Russian line. It must be admitted that Kamimura had no better success than his subordinates in bringing to battle the elusive foe, who was intent more on evasion than combat.

Visual contact was regained at 1755; the Russians came into view to the northwest, on Togo's port bow, steering for Vladivostok. Togo turned in to close the range, but heavy Russian fire and the fact that the setting sun blinded his gunners from behind the Russian line obliged him to sheer off once more. He opened the range by 2000 yards, but maintained visual contact, knowing that with sunset, the light would favor his gunners. So it proved; by 1800 the sun was down and the Russians were silhouetted against the afterglow. *Alexander*, still in the lead, bore the brunt of the fire as the Japanese made good shooting. By 1815 *Alexander*, thoroughly aflame, her bows little more than a tangle of broken steel, turned to port out of control and vanished into the gathering dusk. Kamimura's cruisers saw her capsize and sink at around 1900. The question of survivors is a matter of controversy.

There were none, says Busch; 4 according to Potter and Nimitz; while the Warners give 60. I incline toward the lower estimates, since the only account of *Alexander*'s end is the circumstantial report of the Second Division.

Japanese fire had meanwhile shifted to *Borodino*, which battleship now held the unenviable spot at the head of the line. Togo, unwilling to risk his ships to the hazard of night battle, fairly quickly ordered a withdrawal. Turning away, *Fuji* fired a last round from her forward 12-inch turret; it found the mark, entering *Borodino* near the foremost secondary turret. Two of her magazines detonated, the explosion quite spectacular in the gloom, and she capsized and sank in three minutes. One seaman of her company survived to be taken prisoner from the water the next morning.

Nikolai was technically the fleet flagship, and had been since the loss of *Suvarov* and *Osliabia*, but Admiral Nebogatov had declined to take the initiative until now. As the Japanese turned away to leave the night clear for their torpedo craft, he made "Follow me" and led away to the southwestward. There were two schools of thought in the Russian squadron concerning defense against night torpedo attack. Nebogatov was the foremost exponent of the theory that the ships ought to ride blacked out and fire only if fired upon while others held that it was more effective to make a constant illumination by searchlights and to fire at anything resembling movement. In the course of the night, the former course was to prove correct.

The aggressive defense advocates fared thus: *Navarin*, already damaged by shellfire, took a torpedo aft before succumbing to two of a string of twenty-four mines laid across her bows. *Dmitri Donskoi* survived a destroyer attack. *Nakhimov* and *Monomakh* were torpedoed, but both survived the night *Sissoi Veliki* was torpedoed right aft, but kept afloat without difficulty. On the other hand *TB34*, *TB35* and *TB69* were sunk during attacks. Of those ships that blacked out, none was detected during the night, although two torpedo attacks were successfully fought off before dark.

Dawn of May 28 found Admiral Nebogatov's flagship *Nikolai* with only *Orel*, *Seniavin*, *Aprakain* and *Izumrud* still in company. *Orel*, his only nominally battle worthy unit, was in sorry state, with two of her four 12-inch guns destroyed and both her masts down. Promptly, at

0500, Japanese light cruisers sighted the five ships and reported their presence and position to Admiral Togo, who was marking time 120,000 yards to the north. The Japanese battle fleet approached, visual contact being made first with the First Division plus *Asama*, and subsequently with the remainder of the Second Division. At 1115 the Japanese opened fire at 12,000 yards, beyond the range of Russian guns; Nebogatov accordingly determined to surrender, rather than attempt to close the range. The international code signal "We surrender" brought no slackening of Japanese fire; neither did the raising of a white tablecloth in lieu of a flag, nor even the hoisting of the Japanese ensign. It was only when the Russians had stopped engines that firing ceased; in the meantime the speedy *Izumrud* had made her escape through a gap in the closing Japanese encirclement. (She later ran aground north of Vladivostok and became a total loss.) The remaining four ships were taken without further incident.

The separated Russian ships fared as follows: *Sissoi Veliki* and *Nakhimov* scuttled around dawn to avoid capture. *Monomakh*, approached by a Japanese cruiser, lowered her flag, opened her valves, and went to the bottom. *Svetlana*, encountering *Niitaka* and *Otowa*, was sunk by gunfire after a hotly-contested running engagement.

Donskoi had been in company with *Buini,* but the destroyer having exhausted her fuel, the cruiser sank her by gunfire. Subsequently *Donskoi* encountered the two cruisers that had sunk *Svetlana*, in company with the four cruisers of the indefatigable Admiral Dewa. *Donskoi*, sustaining heavy damage, nevertheless stood off her foes from 1730 until dark, running at a small bay in Matsushima Island, where she was scuttled on the morning of May 29.

In the meantime, *Ushakov*, coming up far astern of Nebogatov's squadron, was intercepted by *Yakumo* and *Iwate* and advised to surrender. This, the obsolescent "flat-iron" declined to do, and the two armored cruisers sent her to the bottom with 85 men after a half-hour's gun battle.

For the rest, *Blestyashtchi, Buistri, Beguprechni* and *Gromki* were sunk by Japanese forces; *Oleg, Aurora* and *Jemtchug* reached Manila and were interned; *Almaz, Grozni* and *Bravi* alone attained their objective, entering the harbor at Vladivostok. Russian losses totaled

4850 men killed and around 7000 prisoners. Japanese casualties were 110 killed and 590 wounded.

The captured units were refitted, partially rearmed, and incorporated into the Imperial Japanese Navy under the following: names:

Orel/Iwami,
Nikolai/Iki,
Seniavin/Mishima,
Apraksin/Okinoshima and
Biedovi/Satsuki.

CONCLUSIONS REGARDING THE RUSSO-JAPANESE WAR

The torpedo continued to disappoint its advocates during the course of this conflict, missing the target far more often than it hit, and failing to sink those ships that it did strike. Indeed, the mobile underwater weapon was not to come into its own until World War I, and then not in naval combat proper, but launched by submarines against merchant ships.

The mine, on the other hand, making its debut in its modern form, was a striking success, generally sinking those ships that ran upon it, and exerting considerable influence upon the maneuvers of both combatants (although it must be admitted that this moral effect was shared by the torpedo).

The major surprise for the experts, however, was the great range at which gun combats occurred. It had been believed that the maximum distance for effective gunnery would be about 3000 yards, with optimum range being between 500 to 1000 yards. In fact, as we have seen, range seldom fell so low as 3000 yards, with engagements taking place at as much as 14,000 yards. At these ranges, the difficulties in spotting the fall of shot for individual guns were very great. It was clear that a system of centralized fire control, already under independent development in the United States and Great Britain was essential, and its adoption was accelerated. This in turn was conducive to the development of the "all big gun" warship, of which the prototype, HMS *Dreadnought*, was already under construction by the war's end. If, however, the desirability of the *Dreadnought* is to be called a lesson of Tsushima, it must be pointed out that it was a lesson not learned at once by the victors, for Japan proceeded to lay down a further series of mixed-armament battleships and armored cruisers. It is not uncommon that the victors have an undue confidence in current weapons, more than other observers may, having obtained victory with them. Consider the United States continued investment in the aircraft carrier following World War II while most other nations pursued a different course in armament. The US Navy still has more aircraft carriers then the rest of the world's navies combined. This seems reminiscent of the "two power standard" policy of the United Kingdom in the 1880's. (see page 12). The United Kingdom had the world's largest navy during this era, but it saw little action, and fought in no major battles. The United States has likewise had the world's largest navy since World War II, but it has seen little action except in the role of supporting land operations. The aircraft carrier now sees itself in the role of supporting land forces and bombarding enemy land based forces, quite similar to the role of the battleship in the second half of the twentieth century. If the reader will allow me: this is a verification of George Santayana's observation on repeating history when it is not studied and understood.

As regards the tactics of the various encounters, there is little enough to be added. The supine inertia of the Russian leadership rendered Japanese victory inevitable. Admiral Togo, long all but deified in Japanese naval circles, proved a competent commander, although his persistence in making long turns when time was of the essence might have compromised his effectiveness against a better-led, better-trained foe.

POSTLOGUE
and
POSTMORTEM

With the Battle of Tsushima the final pieces were in place, or soon would be, for the design of the modern battleship: turrets mounting large guns that were aimed and fired from a central fire control station. These ships were commonly referred to as "all big gun" battleships possessing minimal numbers of smaller calibers intended to be used against torpedo craft not enemy battleships. Battles in the future would be fought with big guns fired in salvos at distant enemies directed from a fire control station. The future battleship would bear the designation of the first such ship to be launched: HMS *Dreadnought*, commissioned by the Royal Navy in 1906. Its construction was begun before the Russo-Japanese War was ended, being laid down on October 2, 1905, a little more than four months after the Battle of Tsushima. Dreadnoughts immediately became the standard warship rendering obsolete all previously built battleships now termed, pre-dreadnoughts and including all the battleships that had fought at Tsushima. Within a few years, still larger and more advanced battleships were being built and the term super-dreadnought was being applied to these.

Even as these ships came to dominate the world's navies, their effect on history was diminishing. Never again would a battle between battleships alter the course of human history as the Battle of Tsushima had done. In World War I the German and British battle fleets would engage each other at the Battle of Jutland, the only full fleet action of the war, and the largest engagement of battleships in history: 55 battleships and battlecruisers on both sides and a total of 250 ships present. The outcome did nothing except continue the status quo and infuse a spirit of caution into the respective navies. Winston Churchill would say: "Jellicoe (the commander of the British fleet) is the only man on either side who could lose the war in an afternoon", and that attitude would dominate the British and German naval commanders for the rest of that war. World War II would see the final eclipse of the battleship although not its disappearance. Two years before the Battle of Tsushima a technologic achievement took place that would bring an end to the battleship as the dominant naval vessel.

On December 17, 1903 two brothers named Orville and Wilbur Wright made a total of four flights in a self-powered airplane, the Wright Flier. The first flight was piloted by Orville, the honor decided by a flip of a coin. It lasted 12 seconds, covered 120 feet and rose only 20 feet above the beach at Kitty Hawk, North Carolina. The longest flight that day was 59 seconds and covered 852 feet. The final landing caused significant damage to the craft, ending the day's activity. The airplane flew straight, no turns or maneuvers were attempted except those necessary to maintain it in flight. The Wright Flier might never have achieved flight at all if it had not been able to take off into a 20 mile per hour wind. (The airplane had a speed of only 7 or 8 miles per hour according to Orville Wright). Only one newspaper, the local journal, made mention of the flight.

Few appreciated the enormous significance of the achievement at the time and no one would have predicted that in less than 40 years (a span of time similar to that covered in this book) the progeny of the Wright Flier would disable the largest Nazi warship, the *Bismarck*, allowing it to be brought to battle and destroyed; would destroy or disable eight battleships of the United States Navy's Pacific Fleet at Pearl Harbor; and would sink the *Prince of Wales*, Britain's most modern battleship and the battlecruiser *Repulse*. These last two vessels, a battleship and a battlecruiser, were the first sunk at sea by unassisted aerial attack. The *Bismarck* was disabled by air attack, but sunk by gunfire and torpedo. All three of these events took place in 1941, only 36 years after the Battle of Tsushima and 38 years after the Wright brother's first flight.

(Of note, none of the eight battleships at Pearl Harbor was technically sunk. The harbor was too shallow to allow these ships to disappear completely below the surface of the water, the navy's definition of "sunk". Two were only slightly damaged, *Pennsylvania* and *Maryland,* and were back in action less than a month after the attack; two were "lost in action", the navy's terminology again, *Oklahoma* and *Arizona*, and were never refloated or repaired. The other four were repaired and improved, and fought in the battles of WWII. They are: *California, Tennessee, Nevada,* and *West Virginia*.)

The last battle between battleships took place on October 24, 1944 at the Battle of Surigao Strait where the US Navy triumphed over the Imperial Japanese Navy. The US battleships included: *California, Maryland, Pennsylvania, Tennessee* and *West Virginia* (survivors of Pearl Harbor) as well as the *Mississippi*. In a night engagement, *West Virginia* opened fire at a range of 22,800 yards using radar-guided fire control. The Japanese were not equipped with this new fire control system and could not reply. Near the end of the battle, *Mississippi* was able to fire for the first time as she also lacked the radar directed fire control system. She fired only once, a full salvo of twelve 14-inch guns. These were last guns ever fired by a battleship at another battleship, fired at the Japanese *Yamashiro*. Both she and *Fuso*, the only other Japanese battleship present, were sunk.

By this time battleships were employed principally as support for amphibious activity. *Nevada,* another of the ships from Pearl Harbor, bombarded the beaches of Normandy during the D-Day landings in 1944. That role would expand even more in the post-World War II era. The last battleship to be decommissioned by the US Navy was the USS *Wisconsin,* which was kept in service because it was argued that it was needed to provide support for marine action on shore. She was finally struck on March 17, 2006.

If this brief account of history is not sufficient to induce a sense of humility in the student of that subject, allow me to raise one further question: What events that are now in our news are predictors of the next weapons? Are we preparing, as so many before us have done, to fight yesterday's war? Is the Internet the next battleground? Our weapons now include computer chips and "smart" technology, but that technology is daily penetrated by "hackers". Will those weapons be used by us or against us? Or are the drugs being brought into our country the weapon being used to conquer us? The cybercrime and drug epidemic we fear so much may be about more than just getting at our money. These may be the weapons in a war to dominate this country and the world, a war where geography no longer matters. Is the next war going to be fought with yet unanticipated weapons on battlefields of our enemy's choosing? Are the aircraft and tanks we arm ourselves with at such great expense as obsolete as the battleships that aircraft rendered useless not so many years ago?

Sun Tzu advised centuries ago in his classic study on this subject, *The Art of War*, that one must not let the enemy choose the battleground. "You can and must choose the ground over which you battle." That advice rings as true today as when it was written.

The *Bismarck* was the ultimate German battleship, heavily armed and armored. During her final battle, she sustained approximately 400 shell hits and at least three torpedoes, one probably launched from HMS *Rodney.* If true, it is the only instance of a battleship successfully torpedoing another battleship. Still, many maintained that *Bismarck* would not have sunk except for the scuttling charges set by the crew as they abandoned ship. The *Bismarck* was totally disabled when abandoned, immobile and afire with no operable armament. Of her crew of more than 2200, only 400 were able to get off her and into the sea and of these only 114 survived. Three were picked out of the sea by a German U-boat and another two by a German trawler. The rest were taken aboard the British cruiser *Dorsetshire* and the destroyer *Maori,* before they broke off rescue efforts when German U-boats were reported in the area. The *Dorsetshire* pulled away with German sailors still hanging onto the ropes lowered over her sides.

The circumstances surrounding the rescue are still controversial, but what is known will give insight into the fate of the men who find their ship sunk from beneath them throughout time. Captain Benjamin Martin was in command of the *Dorsetshire* and was ordered to pick up survivors. While the exact order is unavailable, in another situation (the sinking of the German ship *Scharnhorst* in 1943) the supposed order was to "take a sample" of the survivors. True or not, this phrasing makes it clear that the purpose of the rescue of enemy survivors was not a humanitarian one, at least not exclusively so, but was an intelligence effort: to pick up sailors who might provide information about the enemy navy. It is also important to keep in mind that the primary responsibility of a naval officer is to his command: Captain Martin was responsible for *Dorsetshire*.

Also on the *Dorsetshire* was Midshipman Joe Brooks who saw firsthand the burning and sinking *Bismarck*. He said in a newspaper interview years later that he felt "the utter dismay at the futility of war." When he saw a German sailor whose arms had been blown off, holding onto one of the ropes with his teeth, Brooks recalled: "I could only feel the greatest compassion for him." another sailor, like Brooks himself. Brooks jumped over the side of *Dorsetshire* and tried to tie a bowline around the man but could not and the German was lost. When Brooks was taken back aboard *Dorsetshire* he was confined to his quarters on the grounds that he had deserted his post in the presence of the enemy. "Jesus", was Brooks' response.

Brooks went on to achieve the rank of Lieutenant Commander and was a volunteer in British special operations, midget submarine section, referred to as "X-men" because the craft were "X-class boats". He was awarded the Distinguished Service Cross for his bravery in action in 1944.

It is important to remember that while this story is about the ships, it is also about the people who served on those ships: who exhibited extraordinary courage and gallantry, and without them and their actions, the ships would mean nothing. I think Joe Brooks would have agreed. His obituary and other details of his life are contained in a web link cited in the bibliography.

The wreckage of the *Bismarck* was discovered and explored in 1989 by Dr. Robert Ballard, the discoverer of the *Titanic*. His documentation helped to verify much of what is known and supposed about the *Bismarck*'s fate.

Of the other large battleships sunk in WWII, the Japanese *Yamato* was located in 1982 and there is discussion of raising her from 1125 feet. Her sister, *Musashi* was located in March of 2015 by Microsoft co-founder and philanthropist, Paul Allen and his team, using a remotely controlled robotic vehicle in 3300 feet of water.

While several other wrecks are available to scuba divers, many from this time period, these three ships lie in water too deep for the casual, curious scuba diver to reach, which seems to me as it should be.

"The sea will grant each man new hope, his sleep brings dreams of home." Christopher Columbus.

A Short List of Vessels and Museum Ships Worthy of Visiting

The vessels listed here are still intact much as they were when they were the pride of the navies in which they served. Some, like HMS *Victory* and USS *Constitution*, still are the pride of their nations. Both of these are still fully commissioned vessels. *Victory* is the flagship of Britain's First Lord of the Admiralty. Others, like *Mikasa,* are the only surviving representatives of their kind. HMS *Warrior* represents the first of her kind, the first iron hulled ironclad and begins the story written here. The remaining: *Huáscar, Olympia* and *Aurora* deserve their place by the position they have in the history of naval conflict and warship design, not to mention the outstanding and intense individual history all these vessels share.

I have not listed all vessels that are available, but only those that I personally find most interesting and of course most available. All of these faced threatened destruction at the hands of the nations that now revere them and all survived, sometimes by total serendipity or by individual intervention. All have benefited from the efforts of those who understand the importance of preserving our past in order to assure our future. The story of these valiant examples of perseverance are worthy of telling themselves, but space prohibits such a recording here. That lack of record in no way diminishes those efforts and makes the results all the more impressive.

The specifics of these vessels are assumed to be their current or last recorded statistics and may vary from the information contained in the ships tables.

HMS *Victory*

Length: 227' oa, 186' gundeck
Construction started: July 23, 1759
Launched: May 7, 1765
Weight: 3,556 tons
Armament: (at Trafalgar) Gundeck: 30 32-pdr Blomefield
Middle gundeck: 28 24-pdr
Upper gundeck: 30 12-pdr
Quarterdeck: 12 12-pdr
Forecastle: 2 12-pdr, 2 68-pdr carronade
Place built: Chatham, United Kingdom

The *Victory* is the only surviving first rate Ship-of-the-line and is famous as Nelson's flagship at the climactic Battle of Trafalgar. She is restored and available to be toured. In addition to her historic importance she is an outstanding example of the warships of her era.

Victory is listed as a 104 gun ship justifying her classification as a First Rate Ship-of-the-line, able to stand in line of battle. Her armament varied somewhat during her career. Her size was limited by the material from which she was made, timber, and specifically the size of trees, old growth timber that was available, particularly for her masts. It was necessary to import lumbar to Great Britain at the time of her construction since trees large enough were scarce. It is estimated that 6000 trees were used in her construction, mostly oak. While the adoption of iron and steel ships by the worlds navies had many causes, one that is often overlooked is that the raw material necessary for the construction of wooden ship-of-the-line, old growth timber, was largely depleted by the 19th century.

Also noteworthy is that *Victory* was launched 40 years before her most famous battle, a battle in which she was the premier vessel, a testimony to her durability. Ships of this era lasted considerably longer than the iron ships discussed in this volume that were obsolete within a few years. She had an extensive refitting in 1800 to 1803 which involved adding additional gun ports.

211

She was saved after years of neglect by the intervention of two kings, Edward V and George V. Perhaps apocryphal, it is said that when Thomas Hardy, who was First Sea Lord, returned home one evening and told his wife he had signed the order to scrape the *Victory*, she burst into tears and sent him back to rescind the order. While this tale may be a romantic fabrication, it is also said that page in the official ledger for the date in question has been torn out. *Victory* is the oldest commissioned warship and the flagship of the First Sea Lord since 2012.

Victory may be toured at the National Museum of the Royal Navy, HM Naval Base (PP66), Portsmouth, United Kingdom, PO1 3NH.

http://www.hms-*Victory* .com

USS *Constitution*

Ordered: 1 March 1794
Launched: 21 October 1797 at a total cost of $302,718.00
Nickname: "Old Ironsides"
Status: In active service
Type: 44-gun frigate
Displacement: 2,200 tons
Length: 304 ft bow spirit to spanker
Armament: 30 24-pdr long gun
20 32-pdr carronade
2 24-pdr bow chasers

The *Constitution* is perhaps the most revered vessel in the US Navy, this in a nation whose navy has had many vessels worthy of honor. She was built as part of the new nation's first sea going navy by Edmund Hartt, one of the many talented shipbuilders in the United States,. The story of these first ships of which the *Constitution* is the most famous is well told in Ian W. Toll's excellent book: *Six Frigates: The Epic History of the Founding of the U.S. Navy.*

She served in many roles throughout her history but most notably during the War of 1812 fought between the United States and the United Kingdom. She defeated several smaller British warships and captured much merchant traffic, but her most famous victory was on August 19, 1812, against the 36 gun frigate HMS *Guerriere*. It was during this engagement that she was given the nickname "Old Ironsides". Her sides were not iron of course but oak, still her crew propagated the story that shot from *Guerriere* bounced off her sides as if they were iron. The fame she earned saved her from scrapping and she is still a fully commissioned vessel in the US Navy.

A note about Captain Dacre, commander of the *Guerriere*. He was held in high regard by Captain Hull of *Constitution*, who refused to accept his sword in surrender, acknowledging his bravery and ability during the battle. As was standard practice in the British navy when a ship was lost, he stood trial at court martial and was acquitted completely. The court martial implicated the construction of *Guerriere*, French not British built, and rot present to be at fault.

Constitution is currently undergoing repairs expected to take three years at a cost of $12-15 million in the Charlestown Navy Yard, Boston, MA. Compare this to the original cost, $302,718.00 which I will not adjust for the intervening years, and one can appreciate the value attached to this vessel. I have found no one who has suggested the $15 million should not be spent on her repairs.

https://ussconstitutionmuseum.org/

Warrior

Launched: 29 December 1860,
Commissioned: August 1861
Decommissioned: 31 May 1883
armored frigate
Displacement: 9,137 long tons
Length 420 ft oa
Armament: 26 68-pdr MLSB
10 110-pdr BL
4 40-pdr BL
Armor: 4.5 inch belt and bulkheads
Steam powered with full ship rig

The Crimean war provided the first real naval action since the Napoleonic wars of Admiral Nelson. The French had launched the *Napoléon*, a steam powered vessel, in 1850 establishing steam over sail as propulsion for warships. During the Crimean War (1853 – 1856) a Russian fleet firing explosive shells in the Battle of Sinop (November 30, 1853) completely annihilated the wooden Ottoman fleet opposing it. Then in 1859, France launched the *Gloire*, the first ocean going ironclad. *Warrior* was the British response.

Warrior is the first iron hulled, iron armored warship. Built in 1859-1861 and decommissioned in 1883, she had an active career of only 22 years, so fast were the changes in naval design moving. She is the only vessel listed in this section of ships suggested for touring that never engaged in battle. Even so she is well worth a visit for the opportunity to see this key transition in warship design.

She is berthed as a museum ship with HMS *Victory* at the Portsmouth, UK. National Museum of the Royal Navy, HM Naval Base (PP66), Portsmouth, PO1 3NH.

http://www.hms-Victory .com/

Huáscar

Ordered: 4 August 1864
Builder: Laird Brothers, Birkenhead, England
Launched: 6 October 1865
Commissioned: 8 November 1866
Displacement: 1,900 tons
Length: 219 ft oa
Installed power: 1,650 hp,
Sail plan: Brig-rigged
Armament: (current)
> 2 10-inch Armstrong guns in a single turret
> 2 4.7-inch Armstrong guns
> 1 12-pdr
> 1 0.44-cal Gatling gun

Huáscar is one of the few surviving examples of warships from the early, ironclad era. If one includes the brief time she was in the possession of rebel forces in 1877, she served in three navies, the Peruvian, the Chilean and the rebels supporting Nicolas Pierola. She was built in the United Kingdom and named for one of the last kings of the Incas, possibly because she was built at a time when Spain was attempting to reassert control of Peru. She is much revered by both Peru and Chili being considered the "grave" for both Admiral Miguel Grau of Peru, and Captain Arturo Prat of Chile, who died attempting to board and capture her.

She is often criticized for her design and appearance with the main battery being mounted on the central main deck, and is sometimes referred to as a "monitor". This is an assessment that fails to take into account the restriction on the design of iron ships. Iron vessels were restricted to 300 feet in length thus making them unstable when top heavy. The iron guns and their mounts of this era, particularly turret mounts, were too heavy to be placed above the main deck without making the vessel prone to capsize. Many British ships share this design.

Huáscar is now a museum ship berthed in Concepcion, Chile.

Olympia

Length 344'1" oa
Beam: 53 feet
Displacement: 5,865 tons
Top Speed: 20 knots
Main armament: 4 8-inch/35 in twin turrets
                    10 5-inch/40 QF
                    1 4.6-pdr QF
                    6 1-pdr QF
                    6 18"tt

Launched in 1892, the cruiser *Olympia* (C-6) is the oldest steel warship afloat. Her only rival for this distinction is *Mikasa*, also of steel construction but launched in 1902. The *Warrior* and *Huascar* are older, being launched in 1861 and 1866, but they are iron not steel construction. This distinction is discussed at length in the text.

The pillbox turrets and barbettes are characteristic of ship design at the time they were built, particularly in the US Navy. The height of the funnels and masts are characteristic as well. Engines were coal fired and relied on the funnel to provide draft into the firebox and therefore the efficiency of the engines. The masts were used for communication and observation and height was necessary for both. No canvas was ever intended.

Olympia is available for tours along with several other attractions making a trip to the Philadelphia Shipyard well worth the effort. The 1944 submarine *Becona* is open for touring. She served in the Pacific during World War II and then during the cold war years. She represents the most advanced design in the pre-nuclear US Navy and is representative of the submarines built there. Currently under construction is a full size waterline replica of the American topsail schooner, *Diligence* of 1797.

Olympia is a ship of great historic significance and worthy of a visit for this reason alone, but she is also an excellent example of the culmination of warship design at the end of the 19th century.

The *Olympia* was decommissioned in 1922 and has been part of Independence Seaport Museum since 1996. She is a National Historic Landmark, a National Historic Mechanical Engineering Landmark, is on the National Register of Historic Places, and is part of the Save America's Treasures program.

    Olympia is located at the
Independence Seaport Museum,
211 S. Columbus Blvd.. Philadelphia, PA 19106

    Activities are scheduled throughout the year and more information is available at:

http://www.phillyseaport.org/visitor-information

Aurora

Builder: Admiralty Shipyard, St. Petersburg, Russia
Commissioned: 29 July 1903
Pallada-class protected cruiser
Displacement: 6,731 tonnes
Length: 416 ft
Armament present now, installed in 1917
  14 6-inch
  4 3- pdr
  3 torpedo tubes (two underwater)

Aurora was a protected cruiser in the Baltic Fleet and subsequently in the Second Pacific Squadron of the Russian navy during the Russo-Japanese war. She survived the climactic Battle of Tsushima and was eventually interned in Manila, then under the control of the United States. She rejoined the Russian Navy in the Baltic Sea and served during World War I, but she is best known and celebrated in Russia for her part in the October Revolution. A shot from her forecastle is reported to have been the signal for the Bolshevik assault on the winter palace.

She was refitted in 1917 to bring her up to the standards of the WWI navies, and served through WWII. She was "sunk aground" at her pier during the siege of Leningrad (1941 – 1944), but was never abandoned by her crew who continued to fight on, adding her guns to the defense of the city. Plans are underway to overhaul her and recommission her as the flagship of the Russian Navy. As she stands now she is the oldest ship in that navy and has been visited by millions, She will reportedly be available in 2016 for viewing once again in St. Petersburg, Russia. Despite her service and refits, she is still a good example of the protected cruiser design, and is well worth seeing for her history alone.

She is now a Museum ship in St Petersburg, Russia.

Petrogradskaya embankment, St Petersburg, Russia, 197046
http://www.aurora.org.ru/eng/index.php

Mikasa

Displacement 15,140 tons
Length 432' oa
Speed 18kt
Main armament   4 12-inch/40   14 6-inch/40
                  20 12-pdr   8 3-pdr   4 2.5-pdr
                  4 18"tt

Mikasa Park

Yokosuka, Kanagawa, Japan Near Tokyo.

Mikasa is a pre-dreadnought battleship, and served as the flagship of Admiral Togo's fleet during the Russo-Japanese War. She is worthy of an inspection on either account. She enjoys a prominent position in the history of Japan although she suffered considerable neglect after World War II.

Mikasa Park itself is quite beautiful by all accounts (I have not visited it myself) with a colorful display of fountains and an impressive peace arch. The park includes a statue of Admiral Heihachirō Togo. The accompanying museum has exhibits and memorabilia regarding the admiral and the war, which established Japan as a major naval power. Also included is a shell from the battleship *Yamato,* at 65,027 tons (71,659 tons fully loaded) the largest battleship ever built. She was commissioned on December 16, 1941. Ironically, just 9 days earlier, the Japanese attack on Pearl Harbor rendered her obsolete. She was sunk by carrier-based aircraft as was her sister *Musashi*. The third member of this class, *Shinano*, was converted to an aircraft carrier. She is the largest warship to be sunk by a submarine, the USS *Archerfish*.

The figurehead piece from the cruiser *Izumi* is also preserved here. *Izumi* was formerly a Chilean warship, *Esmeralda*, built in the United Kingdom and sold to Japan. She served Chile in its pacific wars and Japan in the Russo-Japanese War. The most impressive part of the park, however, is the *Mikasa* herself. She is the only pre-dreadnought battleship still surviving intact and represents the epitome of warships prior to the launch of HMS *Dreadnought* that ushered in the new phase of naval design.

A 10 inch shell
taken from the
USS *Maine*

A 10 inch shell taken from the USS *Maine* where it still lies in Havana Harbor after it exploded on February 15, 1898, the event that precipitated the Spanish American War. Paul's lovely wife, Mary, has placed her hand on the shell to provide some scale for the picture. This projectile is typical of the ammunition used in large naval guns (greater than 6 inches). This type of shell would have a separate propellant charge packaged in silk bags as opposed to the QF (quick fire) shells that would have propellant in a brass casing with the shell. The QF guns had a faster rate of fire, although with advances in loading for larger guns the difference was less significant. Standard shell and bag propellant had less weight and took less space in the magazine. For smaller guns, QF still gave a faster rate of fire and the weight and space issues were less important. The discarded shell casings for smaller QF guns did not interfere with the gun crew, but the larger ammunition casings rolling about could become a serious hindrance.

There is controversy about the rating of the *Maine*. In the US Navy she is usually listed as a 2nd class battleship, while in European catalogs she is listed as a cruiser.

This shell is outside the Everhart Museum that has no other naval exhibits, but does have an excellent display of art and of preserved birds including the now extinct passenger pigeon and ivory-billed woodpecker.

It is located at: 1901 Mulberry St, Scranton, PA 18510.

http://everhart-museum.org/

Below is a scene from the Italian cruiser *Francesco Ferruccio* showing the loading of one of her casement mounted, 6-inch guns. The gun is a breech-loader with the sliding block mechanism. The sailor on the left is probably the gunner, responsible for aiming and firing the weapon. Compare the size of this shell to the 10-inch shell from the *Maine* pictured in the previous photo.

APPENDIX A

A Brief Discussion of Ship Classification

Ironclad: The original armored vessel which was essentially a wooden vessel with iron plate or other armor fixed to the vessel particularly above the waterline. Early examples were of wooden construction with iron plates applied, but were otherwise similar to the wooden, broadside warships they replaced. These are often referred to as ironclad broadsides.

Battleship: Large armored vessels mounting heavy guns as their main armament. The term acquired its current meaning in the 1880's and is used in this book to denote the largest of warships during this period.

Cruiser: A type of warship originally designed to have speed and range to allow it to operate in the growing European Colonies. As the name implies they could travel to distant areas where naval presence was needed. Lighter armor and guns permitted this. These ships were armored and armed sufficiently to face all but the most powerful foes and fast enough to escape the larger enemies, battleships. It is also worth noting here that this was a period of rapid change and many believed that variations of the cruiser could stand against battleships.

Armored Cruiser: As cruisers became more heavily armored two variations emerged, the armored cruiser being more prevalent. These vessels had armor over vital areas, engines, magazines and guns; allowing them to stand against more powerful vessels. At this same time the quick fire weapons became common which were felt to offset their smaller caliber with a significant increase in rate of fire.

Protected Cruiser, the other variation, was a vessel that proved to be the answer to design problems posed by increased demands for speed and range at the same time that increases in shell design required more protection. Protected cruisers had an armored deck sloping down below the waterline covering the vital portions of the ship. Interestingly the Chilean *Esmerelda* launched in 1884, was considered the prototype vessel of this class, Built in the United Kingdom by Armstrong she was in keeping with British policy to allow

223

experimental designs to be tried by other navies first and then adopted for the Royal Navy, usually with improvements and of course in greater numbers. Battleships and cruisers would eventually acquire definitions based on size of guns and displacement and these were used in the naval limitation treaties drafted between the World Wars.

Torpedo Boats: Designed to deliver torpedoes of various types from spar and towed devices to the Whitehead locomotive type as discussed in the text. These vessels were essentially unarmored and only lightly armed except for their torpedoes. They relied upon their speed to penetrate the defenses of larger ships, but were generally disappointing in the period under discussion here. They were fast but had limited range due to fuel capacity and consumption.

Destroyers: Originally "Torpedo Boat Destroyers", were vessels built shortly after torpedo boats began to populate the world's navies. They were larger and more heavily armed than their prey and were designed to destroy the torpedo boats attacking battle fleets. There were intermediate designs discussed in the text such as the Torpedo Gunboat.

<u>Smaller vessels with more flexible definitions.</u>

Corvette: A borrowed name from the previous navies of sailing ships. Corvettes were the smallest vessels to be given a rating in navies, smaller than frigates, but larger then coastal vessels. In this period they filled a similar, although less numerous position. They seldom accompanied battle fleets.

Sloops: Again a borrowed term, smaller still than corvettes. Both vessels were armed with guns and occasionally torpedoes.

Gunboat: small fast vessels with limited range, usually limited to coastal defense. These vessels usually carried a significant armament of quick fire guns, and possibly torpedoes as well. Their guns were their principle weapon.

Various specialized craft began to appear towards the end of this time including mine layers and sweepers. Merchant vessels were occasionally armed or used as blockading ships in harbor entrances as well. These are discussed as they appear in the battles.

Composite: the term "composite" used here refers to an iron or steel framed hull covered with wood. This was somewhat successful for commercial shipping, but had obvious deficiencies when applied to warships.

A Word About Ship Names

The naming of ships varied from nation to nation and to the particular times when names were assigned. I will explore this briefly here for each nation included.

The British had enjoyed the primary position in the world's navies for some time when this book opens and they continued in that position throughout the period that is dealt with here. It is therefore fitting that their naming pattern be addressed first. The heritage of the navy was a strong influence throughout Victorian England and the naming of vessels reflected this. Vessels of the sailing navy, Nelson's navy, had saved England, and they believed saved the world, from Bonaparte. Those vessels and their names were honored almost as much as was Nelson himself, and were therefore handed down to the new navy and would continue to be handed down.

Names such as *Inflexible* or *Warrior* were given to ships to honor their predecessors and to inspire emulation in their progeny. These were the major vessels, the battleships, of Her Majesty's Navy. Lesser ships were named for cities or historic and mythical figures These were generally terms of military strength, but there were ships named *Acorn* and *Acute* alongside *Achilles* and *Ajax*. Towards the end of this period monarchs were designated as well. Queen Victoria had a battleship named for her.

European nations followed the British example in christening vessels as they did in their ships design. Spain, Italy and Austria were fond of monarchs as well and *Re d'Italia*, and *Fredrick Max* faced each other at Lissa, while the United States faced Spaniards aboard *Don Juan de Austria* etc. Spain also had vessels named for their imperial possessions, *Isla de Cuba* and the like which allowed historians to bear witness to the sinking of the vessel and liberation of the colony of the same name, simultaneously. France and Russia followed a similar pattern in the naming of vessels.

Ships in South American navies were generally named for heroes such as *Huáscar:* one of the last Inca sovereigns, as well as more modern personalities. This allows the reader to observe Chilean vessels manned and commanded by speakers of Spanish aboard ships named *O'Higgins* or *Almirante Cochrane,* persons who contributed to the liberation movement that employed several ex-patriots following the Napoleonic wars in Europe. The liberation may well have been a byproduct of the European war. Admiral Lord Thomas Cochrane, 10th Earl of Dundonald, for instance, was very successful in Britain's fight against the French but was dismissed from His Majesty's Navy after a conviction for stock fraud. His conviction mattered not in the least to Chile and the other burgeoning nations in Spain's colonial empire, whom he assisted in their fight to gain independence. He had served England in her war against Spain (and France) after all. His reward from Chile was a vessel's moniker: a vessel he would likely have been quite proud of. The nickname given him by his French adversaries was Le Loup des Mers (The Sea Wolf).

If there is some randomness in the above treated nations, Japan showed no such tendency. Major ships, battleships and cruisers, were never named for people, but for shrines or provinces. The names were generally submitted by the minister of war to the Emperor who made the final choice. Some interesting conventions did present themselves, one being the use of Mura after the name of commercial vessels and occasionally warships. The word translates literally to mean circle, and is felt to refer to the enclosed circle a ship forms in the sea, or perhaps the hope that the ship will make a complete circle and return home. Smaller ship might be named *Kasu*mi (Mist) or *Akatsuki* (Daybreak).

Chinese names followed similar conventions, but one further point of interest in this regard will intrigue those wishing to explore the national psyche. This book sees the destruction of the Chinese navy to its completion, many of the ships being captured and incorporated into the Imperial Japanese Navy. This policy was pursued since Japan is almost completely devoid of natural resources, hence iron and steel had a premium value there, not experienced in other nations. The names of Chinese vessels, however, were often retained at least phonetically. The vessel would have a Japanese name that sounded like the name given in China. Perhaps this reflects an appreciation for the worth of the vessels themselves. *Chi Yuan* was renamed *Sai Yen* for example. The words Yuan and Yen are easily recognized as the terms for the currency of each country but in fact both translate to mean round object: a coin or a ship (See Mura above) All of the names given by these nations were of course, in their language and what is given here is the approximate Anglicization of the phonetic equivalent. I have given the most widely accepted spelling, but this limitation should be kept in mind for Chinese and Japanese ships. The Japanese cruiser *Idzuma* is also spelled *Izuma* for example.

Now to the United States where conventions become much simpler. Names prior to this period were given according to British example for qualities to be admired, albeit "American" qualities; the *Constitution* for example, which name was chosen by George Washington. At the beginning of the period discussed in this book, the Confederate States of America adopted the convention of naming their naval vessels after their states: *Virginia*, *Alabama* and so on. When the states were once again united, that convention was continued and battleships were named after states: Thus, *Virginia* and *Alabama* were battleships in the US Navy as well, even though their names were previously given to warships in the rebellious Confederate States and fought against the US Navy.

The *Kearsarge* (BB5) was the only US battleship not named for a state, but instead was names for the sloop of war that sank the CSS *Alabama*. Only the states of Alaska and Hawaii never had battleships named after them, but neither were they states when the US Navy was building battleships.

Cruisers were named for cities, and destroyers and smaller ships for persons and so on. As a side note, I feel obliged to remind the reader that the cruiser, *Brooklyn*, was named for a city in accordance with the above practice since Brooklyn was an independent city until January 1, 1898 when it became a borough of New York City.

An aside to this aside is that an armored cruiser was named *New York* (ACR-2/CA-2), but apparently for the state not the city. She was the sister ship to the *Maine*, but her name was changed to *Saratoga* and then to *Rochester*, presumably to avoid confusion with the battleships named for the state of New York. It will be recalled that the *Maine* was classed as a 2nd class battleship in the US, but was always a cruiser in European listings. New York City has never given her name to a cruiser, although a 36-gun frigate named for the city was built in 1800, and several warships named *New York* (for the city) were laid down but never completed.

APPENDIX B – SHIPS TABLES

Ships tables are only for the battles discussed.

These are complete as much as is possible from reliable sources such as Conway and others. Discrepancies may still exist however.

FIRST LINE OF THE TABLES

The following information is provided in the first line beginning with the ship name. If more than one ship in a class has near identical features and is involved the same conflict, I have attempted to list those in the ship's name positions each ship class having one line with subsequent vessels following. The first is the "class" name ship as the *Indiana* class battleships in the United States fleet, present at the Battle of Santiago de Cuba during the Spanish American War. This is usually the first member of each class. Subsequent listings in this case are *Massachusetts* and *Oregon*, both *Indiana* class battleships with essentially identical features.

Ship names are the anglicized spelling and may vary particularly for Chinese or Japanese vessels. I have tried to choose the most commonly used English spelling and have refrained from giving alternatives in most cases

SECOND LINE OF THE TABLES

Ships designation (or type: Battleship, iron clad etc.) is the standard used for each vessel. When there is a dispute, the European listing is generally given. The USS *Maine* and her sister *New York*, for instance, were classed as battleships by the United States Navy, but as cruisers, either armored or protected in European tables. More detailed discussion of the differences is presented in the text as they tend to vary from year to year. This is discussed in the above section on ship classification.

Next given in the second line in the displacement (weight or technically the weight of water displaced by the ship). This is assumed to be the fully loaded weight. This is standard as the loaded weight determines the depth that must be allowed for a vessel.

The length then follows this with the following abbreviations used:

oa   length overall or the maximum length measured along the hull parallel to the waterline (also used: *LOA*, o/a, o.a.)

pp   length between perpendiculars or length at the waterline between the forward most bow perpendicular and the stern most perpendicular or the rudderstock if no stern post is present. (Also used: p/p, p.p., LPP, LBP or Length BPP)

wl   waterline length, originally loaded waterline length. The length of a ship at the waterline without consideration of length out of the water, bow or stern overhangs and usually ignoring submerged features such as a ram. (Also used: lwl, w/l, w.l. or LWL)

There are designations for length that include the bowsprit in the length but these are not used here.

The final position in the second line gives the maximum rated speed for each vessel in knots (kt) or nautical miles per hour. A nautical mile is 6076.18 feet, compared to the standard statute mile which is 5280 feet. The nautical mile is historically defined as one 60th (one minute of arc) between two lines of latitude. Today it is a derived number rounded to an even number of meters, 1852 m.

Engine size and design, while it is of critical importance in the development of warships, is not generally discussed and only introduced where it is of particular importance. The general progression toward more efficient engines leading to greater speed and longer cruising range is to be anticipated with the newer designs at each stage of development. Essentially all vessels herein had coal fired steam engines; oil would not be introduced until later and diesel or electric were far in the future.

Nomenclature of naval guns and armament

THIRD LINE OF THE TABLES

The third line in these tables (and occasionally a fourth line where necessary), is devoted to armament with the following conventions observed. It is perhaps the most significant information given and is therefore granted a large and more prominent discussion. If it appears confusing, that is because it is confusing.

At the beginning of this book, the older terminology of naval weapons had already begun to be replaced, but this was not yet applied in any completed form. Standard size of muzzle loading smooth bore weapons firing solid shot was based on the weight of the shot, usually a round iron ball. While guns fired a variety of missiles including explosive shells and shrapnel (grape shot, canister or other such antipersonnel loads) they were rated by the weight of the solid iron load. By the end of this volume essentially all weapons fired shells from rifled breech-loading guns. The following designations will be used here.

Muzzle loading smooth bore weapons generally have no designation. An exception is when a weapon is smooth bore in a ship where this is extremely unusual, such as at a time when smooth bore had disappeared from most naval ordinance. *Margues del Duero*, a Spanish gunboat at Manila Bay was armed with muzzle loading rifled guns designated MLR and muzzle loading smooth bore are designated MLSB for emphasis. Otherwise the convention is followed

ML   muzzle loading rifled guns

BL   breech-loading rifled guns. This is omitted in later ships when all guns were breech loaded and rifled

QF   designates guns loaded with single loads containing both shell and propulsive charge. The breech mechanism in these guns is of the sliding block type.

mg   machine gun

These are of three basic types

The standard machine gun as we know it today, was developed by Maxim and others and had an automatic ammunition feed powered by the recoil of the weapon, with individual loads on a belt or other loading mechanism. These weapons are the model for machine guns today and are designated by <u>mg</u> without any addition.

The gravity fed guns came in two basic types

<u>Gatling</u> gun (invented by Dr. Richard Gatling in 1861 reportedly to reduce the size of armies) was a gravity fed gun fired by turning a crank. Bullets were fed from a hopper above the gun.

<u>Nordenfelt</u> (invented by Helge Palmcrantz with production funding by Nordenfelt, 1873) was a gun with multiple barrels mounted in a row (up to 12 barrels and up to 1 inch in diameter) with gravity fed loading for each barrel. It was operated by a back and forth hand lever.

<u>1-pdr mg</u> A special designation used for the 1 pounder machine gun also known as a "pom pom" for the sound made during firing. It fired 1-pound explosive shells, (the smallest explosive projectiles available) at a rate of up to 300 shells per minute with a maximum range of 4500 yards. At this time it was principally an anti-personnel weapon, Later it would become an anti-aircraft weapon.

Size of guns are usually given in inches measured at the breech or by shell weight for smaller weapons.

If the length of the weapon is deemed important it is given in calibers, the number of diameters of the gun contained in its length. A 40 caliber gun has a length from the breech to the muzzle of 40 multiplied by the breech size, and this value is placed after the gun diameter size with a slash (/) between. Thus a 10 inch/40 BL is a 10 inch breech-loading gun that is 400 inches long (40 times 10 inches).

The common designs were also identified by their specific series number as a Mark 6 etc. Many of these designations are omitted as they lack any real importance for the purpose of this text, but this list will provide a reference when it is necessary.

Designation for additional armaments contained on ships listed here.

tt for torpedo tubes designated in inches

mm(s) for mines

When conflicting data exists I have presented what seems to be the most reliable for the battle in which the ship participated. I have occasionally and purposely omitted ships for which there is little or no reliable information particularly those that did not figure prominently in the battles discussed. It is important to bear in mind that armament varied, sometimes significantly, during the life of a ship. Listed here is the armament at the time of the battle as much as available information allows.

One further note: the ships listed in the tables are the ships that are generally listed as present at the engagement. They may not have played a role of importance and if they are not mentioned in the sources I used, they will not be mentioned in the text and therefore not in the index. They are, when possible, listed in the ships table as a matter of completeness. The ships complete name is provided in the tables, which is sometimes significantly longer than the name by which she is referred to in accounts of the battles.

The information provided is best used for comparison of opposing fleets. Battles that have only a few vessels are not included in these tables, but information is presented in the text.

As examples I have listed below the *Victory*, *Constitution*, *Warrior*; and *Iowa* (BB-61 commissioned 1943): a ship-of-the-line, a frigate, an armored frigate and the last class of battleships built for the US Navy. These four vessels are still afloat and can be toured.

More information is presented in **A Short List of Vessels and Museum Ships Worthy of Visiting** except for *Iowa,* which is outside the period discussed here but which is available for touring at Pearl Harbor, HI.

Victory
> First Rate Ship-of-the-line   3500 tons   222'oa   8-9kt
> 30 32-pdr   28 24-pdr   42 12-pdr
> 2 12-pdr cannonade
> Presented is the armament at Trafalgar.

Constitution
> Frigate   2200 tons   207'pp   13kt
> 20 32-pdr cannonade   30 24-pdr   2 24-pdr chasers
> The above is her armament during the War of 1812

Warrior
> Armored Frigate   9287 tons   420'oa   14kt
> 26 68-pdr MLSB   10 110-pdr BL   4 40-pdr BL

Iowa (BB-61, launched 1943)
> Superdreadnought Battleship
>   45,000 tons   887'3" oa   33kt
> 9 16-inch   20 5-inch   80 40-mm   49 20-mm
> Above is the armament in 1943 at her commissioning.
> The 40-mm and 20-mm guns are antiaircraft weapons

Guns listed for the first three vessels are muzzleloaders unless listed otherwise. The designation cannonade refers to a short-barreled gun as compared to the standard guns, which are termed long barreled when a distinction is required. All guns listed above are long barreled unless otherwise designated. Iowa is obviously of a modern vintage.

IRONCLAD TABLES FOR THE BATTLE of LISSA
JULY 20, 1666

ITALIAN SQUADRON

Re d'Italia
      Broadside Ironclad   5610 tons   326'9 1/2" oa 10.8 kt
      6 10-inch   32 6.5-inch

Re di Portogallo
      Broadside Ironclad   5610 tons   326'9"oa   10.8 kt
      2 10-inch   26 6.5-inch

Ancona
      Broadside Ironclad   4155 tons   266'4"oa   13.7kt
      4 8-inch   22 6.5-inch

Regina Maria Pia
      Broadside Ironclad   4200 tons   265'9 1/2"oa 12.9kt
      4 8-inch   22 6.5-inch

Castelfidardo
      Broadside Ironclad   4190 tons   268"4"oa   12.1kt
      4 8-inch   22 6.5-inch

San Martino
      Broadside Ironclad   4200 tons   265'9 1/2"oa 12.6kt
      4 8-inch   22 6.5-inch

Principe di Carignano
      Broadside Ironclad   5450 tons   259'5"pp   10,2kt
      10 8- inch   12 6.5-inch

Formidabile, Terribile
 Broadside Ironclads   2800 tons   207'10"pp   10kt
 4 5-inch   16 6.5-inch

Palestro, Varese
 Coast Defense Ironclads   2165 tons   202'9"wl   8kt
 4 7.9-inch   1 6.5- inch

Affondatore
 Turret Ram   4000 tons   307'9"oa   12kt
 2 9-inch

AUSTRIAN SQUADRON

Erzherzog Ferdinand Max, Habsburg
      Broadside Ironclads   5130 tons   274'9"oa   12.5kt
      16 48-pdr   4 8-pdr   2 3-pdr

Kaiser Max, Prinz Eugen, Don Juan d'Austria
      Broadside Ironclads   3590 tons   252'2"pp   11.4kt
      16 48-pdr   15 24-pdr   1 12-pdr   1 6-pdr

Drache, Salamander
      Broadside Ironclads   2750 tons   206"pp   11kt
      10 48-pdr   18 24-pdr   1 8-pdr   1 4-pdr

Kaiser
      Wooden (unarmored) steam frigate
                    5194 tons   266' oa   12.5kt
    16 60-pdr   74 30-pdr   2 24-pdr

Other wooden vessels were present in both squadrons, but as noted in the text, they took no significant part in the battle. Little data is available on these ships and they are not included in the ships tables.

PARTIAL SHIP TABLES FOR THE SOUTH AMERICAN ENGAGEMENTS 1864-1891

BRITISH

Shah

    Unarmored Iron Frigate   6250 tons   334'   16.2kt
    2 9-inch MLR   16 7-inch MLR   8 64-pdr MLR
    12 mgs   4 torpedo launchers

Amethyst

    Wooden Screw Corvette   1970 tons   220'   13.25kt
    14 64-pdr MLR

PERUVIAN

Independencia

    Broadside Ironclad   3500 tons   215'   12kt
    1 8-inch MLR   1 7-inch MLR   12 70-pdr MLR
    4 30-pdr MLR

Huáscar

    Ironclad Turret Ship   2030 tons   190°pp   12.3kt
    2 10-inch MLR   2 40-pdr

Pilcomayo

    Wooden Gunboat   800 tons   181'9"   11kt
    2 5.8-inch BLR   2 4-inch BLR

Atahualpa, Manco Capac

    monitors   2100 tons   225'oa   8kt
    2 15-inch MLSB

Republica

    Third Class Torpedo Boat   59'   16kt
    2 spar torpedoes

CHILEAN

Esmeralda (launched in 1855)
>Wooden hulled steam corvette 854t   210'pp   8kt
>12 Armstrong 40-pdr BLR   4 Whitworth 40-pdr MLSB

Esmeralda (launched in 1884 sold to Japan in 1894)
>Protected Cruiser   3977 tons   270'wl   18.25kt
>2 10-inch BLR/32   6 6-inch BLR/40   2 6-pdr QF
>5 2-pdr QF   1 Gardner mg

Almirante Cochrane, Blanco Encalada
>Central Battery Ironclads   33?0 tons   210'pp   12.75kt
>6 9-inch MLR   1 4.7-inch   1 9-pdr   1 7-pdr

Abtao
>Composite Gunboat   1600 tons   227'4"   10kt
>1 5.8- inch BLR   4 4.7- inch BLR

Magallanes
>Composite Gunboat   950 tons   196'9"   11kt
>1 7-inch MLR   1 64-pdr MLR   2 4-inch MLR

Almirante Lynch, Almirante Condell
>Torpedo Gunboats   713 tons   230'pp   20.3kt
>5 5-inch QF   4 5-pdr QF   5 14"tt

Janegueo
>Torpedo Boat   35 tons   100'   19kt
>1 mg   2 spar torpedoes

Fresia
>Second Class Torpedo Boat   25 tons   86'   20kt
>1 mg   2 spar torpedoes

Colocolo, Tucapel
>   Third Class Torpedo Boats   5 tons   48'   12kt
>   1 mg   2 14"tt

Guacoldo
>   Third Class Torpedo Boat   59'   16kt
>   2 spar torpedoes

O'Higgins, Chacabuco (specifics are disputed, see page 27)
>   Wooden corvettes
>   3 8.2-inch (150-pounder) BLR

SPANISH IRONCLADS

Numancia
>   Broadside Ironclad   7189 tons   3l5'pp   10kt
>   8 10-inch MLR   7 8-inch MLR   1 7.9-inch BLR
>   8 mgs   2 14"tt

Two of the ships list in the tables for THE SOUTH AMERICAN ENGAGEMENTS served in three navies during their career, *Huáscar* was a Peruvian vessel initially and was briefly a pirate under command of Nicolas Pierola until she was surrendered by that strange and colorful character, to Peru once again. She was later captured by Chile in whose navy she served, ultimately becoming a museum ship in that country. Today she is revered by both nations. She is named for one of the last Incan rulers before the Inca nation was conquered by Spain and her name was never altered when she was captured by and incorporated into the navy of Chile. It is notable that the Inca never processed a navy of any significance,

Esmerelda (launched in 1884 and not to be mistaken for the earlier vessel of the same name launched in 1855) was built in England for Chile where she served until she was sold, first to Ecuador and immediately thereafter to Japan. This was done in order to maintain Chile's treaty obliged neutrality. Indeed her only service to Ecuador was to sail from the Galapagos Islands to Japan, under the Ecuadorean flag. She was renamed *Izumi* by Japan. She was eventually decommissioned but her bow fixtures can be seen at *Mikasa* Park near Tokyo. When she was launched she was the fastest warship afloat and was visited by the future King Edward VII while still in England where she was built. Thus, both of these vessels served in three navies. She is included here as much for her service in the Imperial Japanese Navy as for her service to Chile.

Esmerelda was designed by George Wightwick Rendel who also designed the Rendel boats used with so little success by the Chinese Navy during the Battle of Foochow. *Esmerelda* is considered by many to be the world's first protected cruiser.

SHIP TABLES FOR THE SINO-FRENCH WAR.
1883-1885

FRENCH

Bayard
>   Ironclad Barbette Ship   5915 tons   265'9"w1   14.5kt
>   4 9.4-inch/19   2 7.6-inch   6 5.5-inch   4 3-pdr
>   12 1-pdr revolvers

Triomphante
>   Central Battery Ironclad   4585 tons   258'w1   12.7kt
>   6 9.4-inch/19   1 7.6-inch   6 5.5-inch
>   8 1-pdr revolvers   4 14"tt

Duguay-Trouin
>   Unprotected Cruiser   3479 tons   294'6"w1   l5.5kt
>   5 7.6-inch   5 5.5-inch   10 1-pdr revolvers   2 l4"tt

Villars
>   Unprotected Cruiser   2580 tons   249'3"w1   14.5kt
>   15 5.5-inch   8 1-pdr revolvers

d'Estaing
>   Unprotected Cruiser   2363 tons   268'9"w1   15kt
>   15 5.5-inch   10 1-pdr revolvers

Volta
>   Unprotected Cruiser   1323 tons   259'   12.5kt
>   1 6.4-inch   4 5.5-inch

Lynx, Vipere, Aspic
>   Composite Gunboats 465- 490 tons 144'7"w1   10-11kt
>   2 5.5-inch   2 3.9-inch   24 1-pdr revolvers

T45, T46
>   "27-meter" Torpedo Boats   31 tons   85'4"pp   18kt
>   1 spar torpedo

CHINESE

(other English spellings are common for these vessels)

Yu Yuen
      Screw Frigate   2650 tons   300'oa   12kt
      2 9-inch   24 70-pdr

Yang Wu
      Wooden Sloop   1608 tons   190'4"   13kt
      1 7.5-inch   2 6.3-inch

Chi An, Fei Yuan, Fu Po
      Wooden Sloops   1258 tons   200'   10kt
      1 6.3-inch   4 4.7-inch

Tang Ching
      Composite Sloop   1100 tons   210'   11kt
      1 7-inch   6 4.6-inch

Fu Hsing
      Composite Gunboat   578 tons   169'6"   8kt
      1 6.3-inch   2 4.7-inch

Chen Wei
      Wooden Gunboat   578 tons   160'   10kt
      2 6.3-inch   2 4.7-inch

Chien Sheng, Fu Sheng
      Rendel "Flatiron" Gunboats 256t   87'   8kt
        1 10-inch

SHIP TABLES FOR THE SINO-JAPANESE WAR
1894-1895

CHINESE

Ting Yuen, Chen Yuen
    Battleships   7220 tons   308'oa   15.7kt
    4 12-inch/20   2 5.9-inch/40   3 14"tt

King Yuen, Lai Yuen
    Armored Cruisers   2900 tons   270'4"pp   16kt
    2 8.2-nch/35   2 5.9-inch   4 18"tt

Ping Yuen
    Armored Cruiser   2150 tons   196'10pp   10.5kt
    1 10.2-inch   2 5.9-inch   4 18"tt

Ching Yuen, Chih Yuan
    Protected Cruisers   2300 tons   250'pp   18kt
    5 8.2-inch   2 6-inch   8 57-mm(6pdr)   4 18"tt

Chi Yuan
    Protected Cruiser   2300 tons   236'wl   16.5kt
    2 8.2-inch/35   1 5.9-inch/35   4 3-inch   4 15"tt

Chao Yung, Yang Wei
    Protected Cruisers   1580 tons   210'pp   16.5kt
    2 10-inch   4 4.7-inch   2 2.75-inch

Kuang Chia
    Dispatch Vessel   1296 tons   221'   14.2kt
    1 5.9-inch   4 4.7-inch   6 37-mm(1pdr)

Kuang Ping, Kuang Yi
    Torpedo Gunboats   1000 tons   235'   16.5kt
    3 4.7-inch/40   4 3-pdr   4 14"tt

244

Tsao Chiang
        Composite Gunboat   600 tons   156'8"pp   9kt
        4 6.5-inch

Fu Lung
        First Class Torpedo Boat   120 tons   140'3"pp   24.2kt
        2 1-pdr(37mm)   2 14"tt

Wei Yuan
        Composite Sloop   1100 tons   210'   11kt
        1 6.25-inch Armstrong MLR
        6 4.75-inch/22 Armstrong BLR

JAPANESE

Chiyoda
> Armored Cruiser   2400 tons   310'wl   19kt
> 10 4.7-inch/40   14 3-pdr(57mm)   3 Gatling mgs   4 14"tt

Itsukushima, Hashidate
> Protected Cruisers   4220 tons   301'wl   16.5kt
> 1 12.6-inch/38   11 4.7-inch/42   5 6-pdr(57mm)
> 11 3-pdr   4 14"tt

Matsushima
> Protected Cruiser   4220 tons   301'wl   16.5kt
> 1 12.6-inch/38   12 4.7-inch/42   16 6-pdr
> 6 1-pdr(37mm)   4 14"tt

Yoshino
> Protected Cruiser   4150 tons   360'pp   23kt
> 4 6-inch/40   8 4.7-inch/40   22 3-pdr   5 14"tt

Takachiho, Naniwa
> Protected Cruisers   3650 tons   300'pp   18.5kt
> 2 10.3-inch/35   6 5.9-inch/35   2 6-pdr
> 10 Nordenfeld & 4 Gatling mgs   4 14"tt

Akitsushima
> Protected Cruiser   3100 tons   301'pp   19kt
> 4 6-inch   6 4.7-inch   8 3-pdr   4 14"tt

Fuso
> Central Battery Ironclad   3710 tons   220'pp   13kt
> 4 9.4-inch/20   4 6-inch/50   6 3-inch   11 3-pdr
> 1 Nordenfeld mgs   2 18"tt

Hiei

    Armored Corvette   2200 tons   220'pp   14kt
    5 6.7-inch   6 5.9-inch   4 1-pdr   2 14"tt

Akagi

    Steel Gunboat   610 tons   154'5"pp   12kt
    1 8.2-inch/22   1 5.9-inch

Kotaka
> First Class Torpedo Boat   203 tons   165'pp   19kt
> 4 1-pdr   6 14"tt

TB22, 23
> Second Class Torpedo Boats 85 tons   128'pp   22.5kt
> 2 1-pdr   3 14"tt

TB21
> Second Class Torpedo Boat   79 tons   118'pp   20.7kt
> 2 1-pdr   5 14"tt

TB5 -15, TB16 -91
> Third Class Torpedo Boats   54 tons   114'9"pp   20kt
> 2 1-pdr   2 14"tt

TB15, 20
> Third Class Torpedo Boats   52 tons   111'6"pp   21kt
> 2 1-pdr   2 14"tt

TB1 - 4
> Third Class Torpedo Boats   40t   100'pp   22kt
> 2 1-pdr   3 14"tt

SHIP TABLES FOR THE BATTLE OF MANILA BAY MAY 1 1898

SPAIN

Castilla
> Wooden Cruiser   3289 tons   236'wl   14kt
> 4 5.9-inch   2 4.7-inch   2 87-mm   4 75-mm   10 mgs
> 2 14"tt

Reina Cristina
> Unprotected Cruiser   3042 tons   278'3"   17kt
> 6 6.4-inch   8 6-pdr QF   6 3-pdr QF   5 14"tt

Velasco
> Unprotected Cruiser   1152 tons   210'pp   13kt
> 2 6-inch   2 3-inch   2 mgs

Don Juan de Austria, Don Antonio de Ulloa
> Unprotected Cruisers 1152 tons   210'pp   13kt
> 4 4.7-inch   4 6-pdr QF   1 mg   2 14"tt

Isla de Luzon, Isla de Cuba
> Protected Cruisers   1030 tons   184'10"pp   5.9kt
> 6 4.7-inch   4 6-pdr QF   4 mg   3 14"tt

General Lezo
> Gunboat   515t   160'   11kt
> 2 4.7-inch   1 3.5-inch   2 mg   1 tt

Margues del Duero
> Gunboat   492t   157'5"   10kt
> 1 6.4-inch MLR   2 4.7-inch MLSB   1 mg

UNITED STATES

Olympia
>   Protected Cruiser   5865 tons   344'l"oa   20kt
>   4 8-inch/35   10 5-inch/40 QF   1 4 6-pdr QF
>   6 1-pdr QF   6 18"tt

Baltimore
>   Protected Cruiser   4415 tons   335'oa   19kt
>   4 8-inch/35   6 6-inch/50   4 6-pdr QF   2 3-pdr QF
>   2 1-pdr QF

Boston
>   Protected Cruiser   3189 tons   283'oa   13kt
>   2 8-inch/30   6 6-inch/30   2 6-pdr QF   2 3-pdr QF
>   2 1-pdr QF

Raleigh
>   Protected Cruiser   3185 tons   305'9"oa   19kt
>   1 6-inch/40   10 5-inch/40 QF   8 6-pdr QF   4 1-pdr QF
>   4 18"tt

Concord
>   Patrol Gunboat   1710 tons   244'6"oa   16kt
>   6 6-inch/30   2 6-pdr QF

Petrel
>   Patrol Gunboat   867 tons   188'oa   11.4kt
>   4 6-inch/30   2 3-pdr QF   2 1-pdr QF

SHIP TABLES FOR THE BATTLE OF SANTIAGO de CUBA, JULY 3, 1898

SPAIN

Infanta Maria Teresa, Almirante Oguendo , Vizcaya
> Armored Cruisers   6890 tons   364'oa   20.2kt
> 2 11-inch   10 5.5-inch   8 12-pdr QF   10 3-pdr QF
> 10 mg   8 tt

Cristobal Colon
> Armored Cruiser   6840 tons   366'8"oa   20kt
> 2 8-inch   14 6-inch   10 3-inch   6 6-pdr QF
> 3 14"tt

Furor
> Destroyer   570 tons   220'   28kt
> 2 10-pdr QF   2 6-pdr QF   2 1-pdr mgs   2 14"tt

Pluton
> Destroyer   400 tons   225'   30kt
> 2 14-pdr QF   2 6-pdr QF   2 1-pdr mgs   2 14"tt

UNITED STATES

Iowa
    Battleship   11,410 tons   362'5"oa   15kt
    4 12-inch/35 8 8-inch/35   6 4-inch/40 QF   20 6-pdr QF
    4 1-pdr QF   4 14"tt

Indiana, Massachusetts, Oregon
    Battleships   10,288 tons   350'11"oa   15kt
    4 13-inch/35 8 3-inch/35   4 6-inch/40   20 6-pdr QF
    6 1-pdr QF   6 18"tt

Texas
    Battleship   6135 tons   308'10"oa   17kt
    2 12-inch/35 6 6-inch/35   12 6-pdr QF   6 1-par QF
    3 14"tt

Brooklyn
    Armored Cruiser   9215 tons   402'7"oa   20kt
    8 8-inch/35   12 5-inch/40 QF   12 6-pdr QF   4 1-pdr QF
    5 18"tt

New York
    Armored Cruiser   8200 tons   384'oa   20kt
    6 8-inch/35   12 4-inch/40 QF   8 6-pdr QF   4 1-pdr QF
    3 14"tt

New Orleans
    Protected Cruiser   3769 tons   354'5"oa   20 kt,
    6 6-inch/50   4 4.7-inch/50   10 6-pdr QF   8 1-pdr QF
    3 18" tt.

SHIP TABLES FOR THE RUSSO-JAPANESE WAR
1904-1965

RUSSIAN PACIFIC SQUADRON,

Many of the lesser vessels, particularly those of the Imperial Russian Navy, present difficulties of specific descriptions at the time of the onset of hostilities. All major capital ships are listed but some of these minor and less significant ships are omitted to avoid confusion where descriptions are incomplete or conflicting in order to avoid further confusion with no real benefit. Complete information is available in the references listed.

Tsarevitch

Battleship	12,915 tons	388'9"oa	18.5kt
4 12-inch/40	12 6-inch/45	20 11-pdr	20 3-pdr
4 15"tt	45 mns		

Retvizan

Battleship	12,900 tons	368'8"oa	18kt
4 12-inch/40	12 6-inch/45	20 11-pdr	24 2-pdr
6 15"tt	45 mns		

Peresvet, Pobieda

Battleship	12,685 tons	434'6"oa	18.5kt
4 10-inch/45	11 6-inch/45	20 11-pdr	3 1-pdr
5 15"tt			

Petropavlovsk, Poltava, Sevastopol

Battleship	11,500 tons	369'wl	16.5kct
4 12-inch/40	12 6-inch/45	12 3-pdr	28 1-pdr
6 18"tt	60 mns		

Rossia

Armored Cruiser	13,675tons	480'6"oa	20.2kt
4 8-inch/45	16 6-inch/45	12 11-pdr	20 3-pdr
16 1-pdr	5 15"tt		

Gromoboi
    Armored Cruiser   13,220tons   48l'oa   20kt
    4 8-inch/45   16 6-inch/45   24 11-pdr   4 3-pdr
    4 1-pdr   4 15"tt

Rurik
    Armored Cruiser   11,690tons   435'oa   18.7kt
    4 8-inch/35   16 6-inch/45   6 4.7-inch/45   6 3-pdr
    10 1-pdr   4 l5"tt

Bayan
    Armored Cruiser   7775 tons   449'7"oa   21kt
    2 8-inch/45   8 6-inch/45   16 11-pdr   8 3-pdr
    2 15"tt

Pallada, Diana
    Protected Cruisers   6823 tons   415'8"oa   19kt
    8 6-inch/45   24 11-pdr   8 1-pdr   5 15"tt

Bogatyr
    Protected cruiser   6645 tons   459'8"oa   23kt
    12 6-inch/45   12 11-pdr   8 3-pdr   2 1-pdr
    2 l5"tt

Variag
    Protected Cruiser   6500 tons   425'oa   23.2kt
    12 6-inch/45   12 11-pdr   8 3-pdr   2 1-pdr
    6 15"tt

Askold
    Protected cruiser   5905 tons   437'oa   23.8kt
    12 6-inch/45   12 11-pdr   8 3-pdr   2 1-pdr mg   6 15"tt

Boyarin
    Protected Cruiser   3200 tons   345'wl   22kt
    6 4.7-inch/45   6 3-pdr   4 1-pdr
    5 l5"tt

Novik
      Protected Cruiser   3080 tons   36O'5"wl   25kt
      6 4.7-inch/45   6 3-pdr   5 15"tt

Boevoi
      Destroyer   350 tons   213'wl   27.5kt
      1 11-pdr   5 3-pdr   2 15"tt

Boiki, Burni
      Destroyers   350 tons   210'oa   26kt
      1 11-pdr   5 3-pdr   3 15"tt   18 mines

Bez Strashni, Beoditelnt, Bezposhredni, Bezshumni,
      Destroyers   346 tons   202'7"oa   27kt
      1 11-pdr   5 3-pdr   3 15"tt

Vnimatelno, Vuinoslivi, Vnushitelni, Vlastni, Grozovoi
      Destroyers   3l2 tons   l85'8"oa   26.5kt
      1 11-pdr   5 3-pdr   2 15"tt

Leitenant Burekov
      Destroyer   280 tons   193'7"   33.6kt
      6 3-pdr   2 14"tt

Razyashchi, Rastoropni, Serditi, Smleli, Skori, Statni. Sil'ni
Steregushchi Storozhevoi Strashni, Stroini
      Destroyers   240 tons   190'oa   2?.5kt
      1 11-pdr   3 3-pdr   2 15"tt   12 mm in some

Amur, Yenisei
      Minelayers   3010 tons   300'wl   18kt
      5 11-pdr   7 3-pdr   1 15"tt   500 mm

Gremzashchi, Otvajni
      Armored Gunboats   1854 tons   257'1"oa   14kt
      1 9-inch/35   1 6-inch/35   6 3-pdr   4 1-pdr mgs
      2 15"tt   20 mns

Djigit, Razboinik
    Sloops   1355 tons   207'6"wl   13.5kt
    3 6-inch/23   4 4.2-inch/20 6 1-pdr mgs   1 15"tt

Korietz
    Gunboat   1270 tons   206'wl   13.5kt
    2 8-inch/35   1 6-inch/35   4 4.2-inch/20 4 1-pdr mgs
    1 15"tt

Sivuch, Bohr
    Gunboats   1250 tons   187'6"wl   11.5kt
    1 9-inch/30   1 6-inch/28   6 4.2-inch/20 4 1-pdr mgs

Vsadnik, Gaidamak
    Torpedo Gunboats   432 tons   197'6"oa   22.5kt
    6 3-pdr   3 1-pdr   2 15"tt

RUSSIAN BALTIC FLEET

(SECOND PACIFIC SQUADRON)

Borodino, Imperator Alexander III, Orel, Knyaz Suvarov
      Battleships   13,516 tons   397'oa   17.8kt
      4 12-inch/40 12 6-inch/45 20 11-pdr   20 3-pdr   4 15"tt

Osliabia
      Battleship   12,683 tons   434'6"oa   15.5kt
      4 10-inch/45 11 6-inch/45 20 11-pdr   20 3-pdr
      8 1-pdr   5 15"tt

Sissoi Veliki
      Battleship   10,400 tons   351'10"oa   15.7kt
      4 12-inch/40 6 6-inch/45   12 3-pdr   18 1-pdr
      6 18"tt

Navarin
      Battleship   10,206 tons   357'8"oa   15.5kt
      4 12-inch/35 8 6-inch/35   8 3-pdr   15 1-pdr
      6 15"tt

Imperator Nikolai I
      Battleship   9672 tons   333'6"wl   15.3kt
      2 12-inch/30 4 9-inch/35   8 6-inch/35   10 3-pdr
      8 1-pdr mgs 6 15"tt

Admiral Ushakov, Admiral Seniavin
      Coast Defense Battleships   4125 tons   286'6"oa   16kt
      4 9-inch/45   4 4.7-inch/45   6 3-pdr   10 1-pdr
      6 1-pdr mgs 4 15"tt

General Admiral Graf Apraksin
      Coast Defense Battleship   4126 tons   286'6"oa   16kt
      3 10-inch/45 4 4.7-inch/45 10 3-pdr   12 1-pdr   4 18"tt

Admiral Nakhimov
> Armored Cruiser   8524 tons   333'wl   17kt
> 8 8-inch/35   10 6-inch/35   4 3.4-inch   6 3-pdr
> 4 1-pdr mg   3 15"tt   40 mms

Dmitri Donskoi
> Armored Cruiser   6200 tons   296'8"wl   16.2kt
> 6 6-inch/45   10 4.7-inch/45   6 3-pdr   10 1-pdr
> 10 1-pdr mgs   5 15"tt

Vladimir Monomakh
> Armored Cruiser   5593 tons   296'3"wl   15.2kt
> 5 6-inch/45   10 4.7-inch/45   8 3-pdr   8 1-pdr mgs
> 5 15"tt

Aurora
> Protected Cruisers   6657 tons   415'8"oa   19kt
> 8 6-inch/45   24 11-pdr   8 1-pdr   3 15"tt

Oleg
> Protected Cruiser   6645 tons   439'8"oa   23kt
> 12 6-inch/45   12 11-pdr   8 3-pdr   2 1-pdr
> 2 15"tt

Svetlana
> Protected Cruiser   3862 tons   331'4"wl   21.6kt
> 6 6-inch/45   I0 3-pdr   2 15"tt   20 mms

Izumrud, Jemtchug
> Protected Cruisers   3103 tons   364'oa   24kt
> 6 4.7-inch/45   6 3-pdr   2 1-pdr   5 18"tt

Buini, Bravi, Blestyashtchi, Buistri, Bodri, Byedovi, Bezuprechni, Gromki, Grozni

| Destroyers | 350 tons | 210'oa | 26kt |
| 1 11-pdr | 5 3-pdr | 3 15"tt | 18 mms |

Almaz

| Armed Yacht | 3285 tons | 365'8"oa | 19kt |
| 4 11-pdr | 8 3-pdr | | |

There were also several auxiliary vessels: collier, supply and hospital ships that accompanied this fleet, but took no part in the battle and therefore are omitted in order to minimize confusion. The English spelling of some Russian ship names is also disputed and while I have attempted to choose the most common of these, the reader should be aware that others exist.

JAPANESE IMPERIAL NAVY

Asahi

Battleship	15,200 tons	425'6"oa	18kt
4 12-inch/40	14 6-inch/40	20 12-pdr	6 3-pdr
6 2.5-pdr	4 18"tt		

Mikasa

Battleship	15,140 tons	432'oa	18kt
4 12-inch/40	14 6-inch/40	20 12-pdr	8 3-pdr
4 2.5-pdr	4 18"tt		

Hatsuse

Battleship	15,000 tons	439'9"oa	18kt
4 12-inch/40	14 6-inch/40	20 12-pdr	8 3-pdr
4 2.5-pdr	4 18"tt		

Shikishima

Battleship	14,850 tons	458'oa	18kt
4 12-inch/40	14 6-inch/40	20 12-pdr	6 3-pdr
6 2.5-pdr	5 18"tt		

Fuji, Yashima

Battleships	12,533 tons	412'oa	18kt	
4 12-inch/40	10 6-inch/40	20 3-pdr	4 2.5-pdr	5 18"tt

Idzumo, Iwate

Armored Cruisers	9750 tons	434'oa	20.75kt
4 8-inch/45	14 6-inch/40	12 12-pdr/40	8 2.5-pdr
4 18"tt			

Asama, Tokiwa

Armored Cruisers	9700 tons	442'oa	21.5kt
4 8-inch/45	14 6-inch/40	12 12-pdr/40	7 2.5-pdr
5 18"tt			

Yakumo
   Armored Cruiser   9646 tons   434'oa   20.5kt
   4 8-inch/45   12 6-inch/40   12 12-pdr/40   7 2.5-pdr
   5 18"tt

Adzuma
   Armored Cruiser   9307 tons   452'6"oa   20kt
   4 8-inch/45   12 6-inch/40   12 12-pdr   12 3-pdr
   5 18"tt

Nisshin
   Armored Cruiser   7698 tons   366'6"oa   20kt
   4 8-inch/45   14 6-inch/45   10 3-inch/40   6 3-pdr
   2 mgs   4 18"tt

Kasuga
   Armored Cruiser   7628 tons   366'6"oa   20kt
   1 10-inch/45   2 8-inch/45   14 6-inch/45   10 3-inch/40
   5 3-pdr   2 mg   4 18"tt

Kasagi
   Protected Cruiser   4900 tons   402'oa   22.5kt
   2 8-inch/45   10 4.7-inch/40   12 12-pdr   6 2.5-pdr
   4 18"tt

Chitose
   Protected Cruiser   4760 tons   396'oa   22.5kt
   2 8-inch/45   10 4.7-inch/40   12 12-pdr   6 2.5-pdr
   4 18"tt

Takasago
   Protected Cruiser   4160 tons   38?'6"oa   23.5kt
   2 8-inch/45   10 4.7-inch/40   12 12-pdr   6 2.5-pdr
   5 18'tt

Niitaka, Tsushima
　　　Protected Cruisers　　3366 tons　　334'6"pp　　20kt
　　　6 6-inch/40　　　　　　10 12-pdr　　4 2.5-pdr

Otowa
　　　Protected Cruiser　　3000 tons　　341'oa　　21kt
　　　2 6-inch/50　6 4.7-inch/40　　4 12-pdr　　2 mgs

Akashi
　　　Protected Cruiser　　2756 tons　　295'3"pp　　20kt
　　　2 6-inch/40　6 4.7-inch/40　　10 3-pdr　　4 2.5-pdr
　　　4 mgs　　2 14"tt

Suma
　　　Protected Cruiser　　2657 tons　　306'9"pp　　20kt
　　　2 6-inch/40　6 4.7-inch/40　　10 3-pdr　　4 2.5-pdr
　　　4 mgs　　2 14"tt

Idzumi
　　　Protected Cruiser　　2920 tons　　270'pp　　18.25kt
　　　2 6-inch　6 4.7-inch　　2 6-pdr　　5 2-pdr
　　　2 mgs　　3 18"tt

Asagiri, Harusame, Murusame
　　　Destroyers　　375 tons　　227'　　29kt
　　　2 12-pdr　4 6-pdr　　2 18"tt

Akatsuki, Kasumi
　　　Destroyers　　365 tons　　220'6"　　31kt
　　　2 12-pdr　4 6-pdr　　2 18"tt

Asashio, Shirakumo
　　　Destroyers　　342 tons　　216'　　31kt
　　　2 12-pdr　4 6-pdr　　2 18"tt

Ikazuchi, Akebono, Inazuma, Ohoro, Sazanami
      Destroyers   305 tons   220'9"   31kt
      1 12-pdr   5 6-pdr   2 18"tt

Kagero, Murakumo, Shinonome, Shiranui, Usugumo, Yugiri
      Destroyers   275 tons   208'6"   30kt
      1 12-pdr   5 6-pdr   2 18"tt

APPENDIX C – RELEVANT MAPS

I have included two maps here in an effort to help with the understanding of the material presented in the text. These two maps are for the conflicts in the Pacific Ocean off the coast of South America and for the Yellow Sea and Sea of Japan, site of the Sino-Japanese War and the Russo-Japanese War. I believe that all other battles are easily understood without this aid, but in these two instances, the geography is difficult to visualize without assistance as well as the fact that the conflicts are spread over both geography and time such that an appreciation of the information provided by these maps, simple as they are, greatly enhances the understanding of the events.

As said, the maps are greatly simplified, listing only the necessary information: national boundaries and sites of battles. In the case of the Pacific War, the national boundaries were altered as a result of the Treaty of Ancón, 1879, and while not crucial to the understanding of the conflict, are none-the-less interesting as Bolivia became a landlocked nation as a result. The loss of her coast is still a contentious issue today.

One further point is apparent upon consulting these maps. The naval operations were actually conducted in relatively small areas. In the Pacific War, the area involved the coast of Bolivia, southern Peru and northern Chile, territory that was ultimately ceded to Chile. That the battles were fought in an area that was disputed and ultimately ceded is significant, I think. Bolivia also lost territory to Argentina not shown here.

As regards the Sino-Japanese War and even more so the Russo-Japanese War, a great deal of the conflict took place in Korea and Manchuria (always an area of disputed claims by Japan, China and Russia). Only Weiheiwei (now Weihai) was in China while Vladivostok clearly was Russian. That Korea was the site of many of the engagements was not deemed noteworthy at the time.

The Island of Tsushima has been claimed by Russia and Korea, but has always remained Japanese. When the climatic encounter of the Russo-Japanese War took place in the Straits of Korea, Japan named the battle for Tsushima, a province of Japan, perhaps to assert its sovereignty and its position as a world power.

MAP OF THE PACIFIC COAST OF SOUTH AMERICA

1864 to 1891

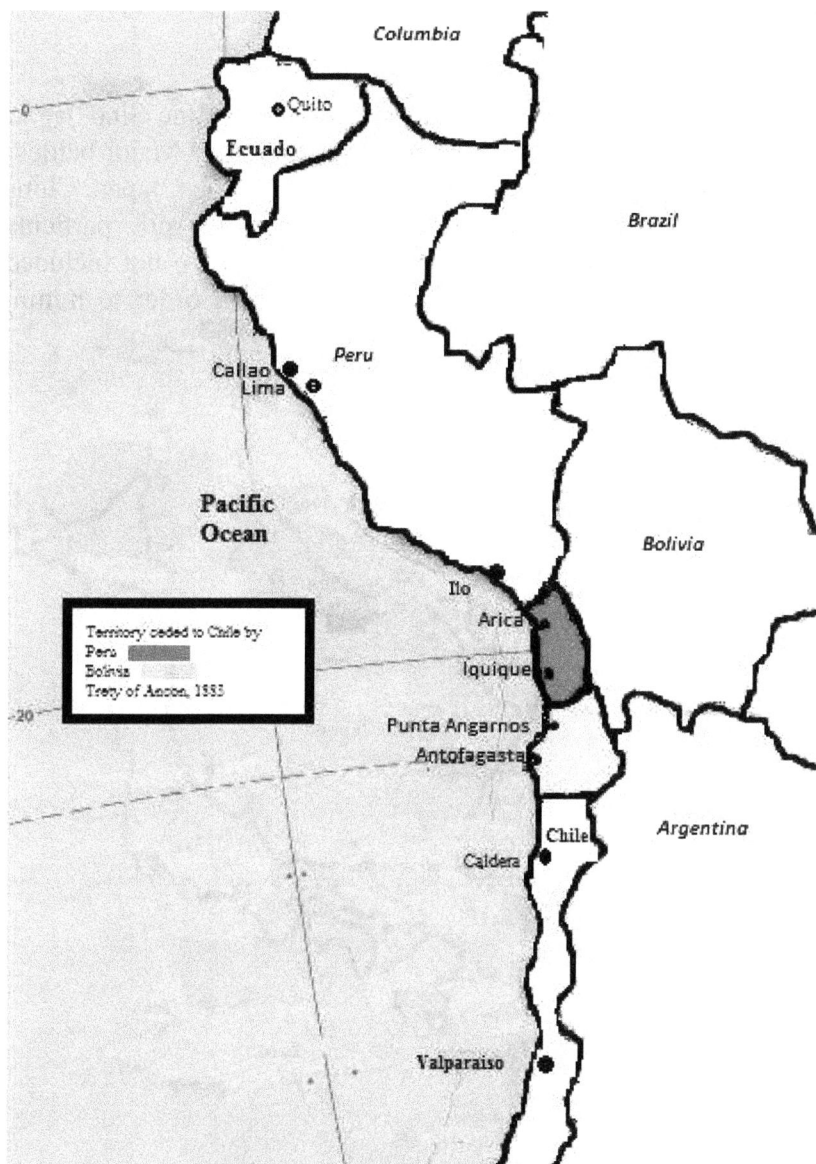

Columbia

Quito

Ecuado

Brazil

Peru

Callao
Lima

Pacific
Ocean

Bolivia

Ilo

Arica

Territory ceded to Chile by
Peru
Bolivia
Trety of Ancon, 1883

Iquique

Punta Angarnos
Antofagasta

Argentina

Chile

Caldera

Valparaiso

MAP OF THE COAST OF ASIA

THE YELLOW SEA AND SEA OF JAPAN

1894 – 1905

The principle areas and nations involved in the Sino-Japanese War and the Russo-Japanese War are indicated here. Major battles are noted, most of which are not in the opposing countries: Japan, China or Russia. There were battles fought on land as well, particularly surrounding Port Arthur and Weiheiwei but these are not included in the text and are therefore omitted on this map in order to minimize confusion.

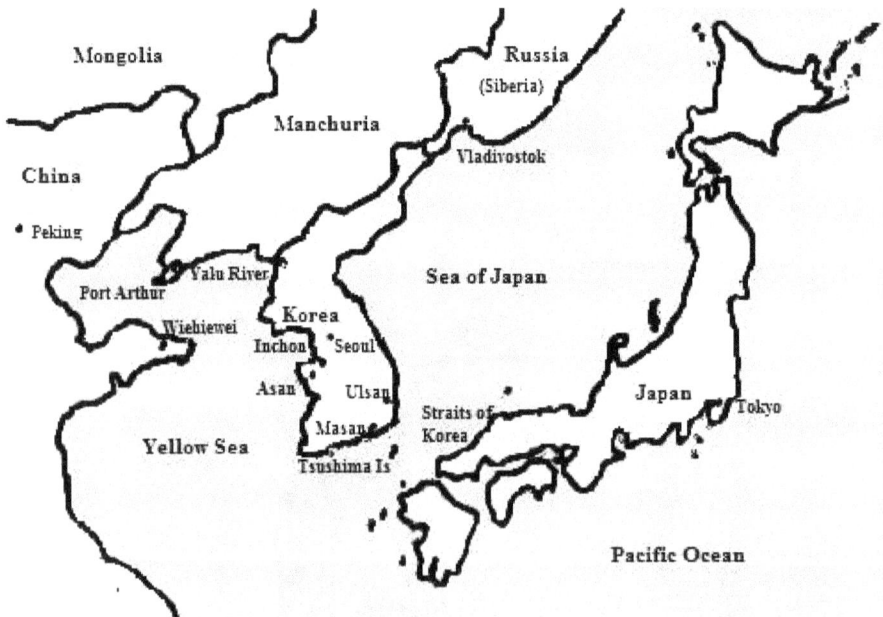

266

SELECTED BIBLIOGRAPHY

Many of the sources listed are books, which were printed decades ago, and many are not available in any other format today. They are therefore listed accordingly, as they were used when preparing this volume. It is my sincere hope that they will not be lost forever. In many cases they contain information available nowhere else.

Akers, Charles Edmond. A history of South America. 1854-1904. New York, E. P. Dutton, 1905

Archibald, James Francis Jewell. War correspondent reporting for Collier's Magazine during the Spanish-American War, The Sino-Japanese War and the Russo Japanese War. His observations can be viewed in that magazine and others. Mr. Archibald was reported to be the first man to be wounded in the Spanish-American War. It was apparently a minor wound, but does speak for his enthusiasm for reporting the "story" firsthand.

Brooks, Joe in: What Ever Happened to Midshipman Joe Brooks, RN? http://nineteenkeys.blogspot.com/2010/05/whatever-happened-to-midshipman-joe.html (This link contains his obituary)

Busch, Noel F. The Emperor's Sword: Japan vs Russia in the Battle of Tsushima Funk & Wagnalls; First American Edition 1969

Conway provides many excellent and detailed volumes regarding ship design and statistics as well as battles. All are a superior source of information, but the most helpful for this period is:
Conway's History of the Ship: Steam, Steel and Shellfire. Greenhill, Basil editor. Conway Maritime Press Ltd, 1992, ISBN 0851776086

Davis, Richard Harding. Captain Philo Norton McGiffin on
http://www.navyandmarine.org/ondeck/1894yalubattle_mcgiffin.htm
    I was unable to find this reference in print

Dupuy, R. Ernest and Dupuy, Trevor N. Harper Encyclopedia of
Military History: From 3500 BC to the Present. Harper Resource; 4
Sub edition, March, 1993 ISBN-13: 978-0062700568

Eastlake, F Warrington and Yoshiaki, Yamada. Heroic Japan : a
history of the war between China & Japan. Ulan Press, August 31,
2012 ISBN-13: 978-1296811112

Emerson, Edwin Jr. and Miller, Marion Mills The Nineteenth Century
and After: A History Year by Year from A.D. 1800 to the Present,
Volume 3, February 4, 2010 ISBN-13: 978-1143735806

Fitzsimons, Bernard Ed. The Illustrated Encyclopedia of 20th Century
Weapons and Warfare. Columbia House, 1977

Farcau, Bruce W. The Ten Cents War: Chile, Peru, and Bolivia in the
War of the Pacific, 1879-1884, Praeger, September 30, 2000.
ISBN 0-275-96925-8

Galdames, Luis. A History of Chili. The University of North Carolina
Press, 1941. Reissued by Russell & Russell, Jan 1, 1964

Gordon, Andrew and Murray, John. The Rules of the Game: Jutland
and British Naval Command. Publisher: John Murray, May 23, 2005.
ISBN-13: 978-0719561313

Hargreave, Reginald, Red Sun Rising: The Siege of Port Arthur. JB.
Lippincott Co. First Edition 1962.

Hood, Christopher Japanese Education Reform: Nakasone's Legacy.
Routledge; 1 edition, April 5, 2001. ISBN-13: 978-0415232838.

Hough, Richard. The Fleet That Had To Die. Ballantine Books, 1960.
    There are several reissues for this title. Above is the one used
    here and below is the most recent.
Hough, Richard.The Fleet That Had To Die. Endeavour Press, April 8,
2015.

Jentschura, Hansgeorg; Jung, Dieter; Mickel, Peter, Warships of the
Imperial Japanese Navy, 1869-1945. Naval Institute Press; 1st edition
November 1976. ISBN-13: 978-0870218934

McElwee, William. The Art of War: Waterloo to Mons Little Hampton
Book Services Ltd., February 27, 1975. ISBN-13: 978-0297768654
    An excellent discussion of the evolution of military theory
    practice an technology.

McGiffin, Lee. Yankee of the Yalu: Philo Norton McGiffin, American
Captain in the Chinese Navy, 1885-1895. E. P. Dutton; First Edition,
1968

The Oxford Companion to Ships and the Sea. Oxford Univ Pr; Book
Club edition, November 1976. ISBN-13: 978-0192115539
The Oxford Companion to Ships and the Sea (Oxford Quick
Reference)
2nd Edition by I. C. B. Dear (Editor), Peter Kemp (Editor), Oxford
University Press; 2 edition, October 23, 2006.
ISBN-13: 978-0199205684
    Both of these are included as references as there are differences
    in the two editions. These differences represent changes in
    available information and changes in emphasis of various
    information and sources. Both are interesting.

Pemsel, Helmut. Atlas of naval warfare: An atlas and chronology of conflict at sea from earliest times to the present day. Arms and Armor Press; 1st edition 1977. ISBN-13: 978-0853683513
Pemsel, Helmut and. Smith, Major G. D. G. A History of War at Sea: An Atlas and Chronology of Conflict at Sea from Earliest Times to the Present. Naval Institute Press, Jun 1979. ISBN-13: 978-0870218033
    Both of these editions were used and are therefore included.

Pike, Frederick B. The Modern History of Peru. Littlehampton Book Services Ltd, June 1967. ISBN-13: 978-0297748519

Potter, E. B and Chester W. Nimitz. Sea Power: A Naval History Bramhall House; First edition, August 3, 1960.

Powell, John Weapons & Warfare. Salem Pr Inc; Har/Psc Re edition February 15, 2010. ISBN-13: 978-1587655944
    There are several editions and updates for this book. Above is the one I found most useful for this work.

Southworth, John Van Duyn. The Age of Steam (War at Sea), Hippocrene Books; 1st Hippocrene edition 1984

Snow, Edward Rowe. Tales of Terror and Tragedy. Dodd Mead; First Edition, December 1979. ISBN-13: 978-0396077756

Swann, Leonard Alexander. John Roach, maritime entrepreneur;: The years as naval contractor, 1862-1886. U.S. Naval Institute 1965

Toll, Ian W. Six Frigates: The Epic History of the Founding of the U.S. Navy. W. W. Norton & Company; Reprint edition, March 17, 2008. ISBN-13: 978-0393330328

Watts, A. J. And B. G. Gordon, The Imperial Japanese Navy. Doubleday, Garden City, NY, 1971.

Warner, Peggy. The Tide at Sunrise: A History of the Russo-Japanese War, 1904-05. Routledge; 1 edition, November 13, 2004 ISBN-13: 978-0714682341

> Peggy and Denis Warner have written several books that deal with the Russo-Japanese War. All are interesting and helpful.

Westwood, J. N. Russia Against Japan, 1904-1905: A New Look at the Russo-Japanese War. State University of New York Press, May 1986. ISBN-13: 978-0887061912

Woods, David J. The Bombardment of Paradise WTA Publishing, Geneva, Switzerland, (July 24, 2011).

Index of Ships, Battles and Persons Cited Here

There are some points that the reader should bear in mind when using this index.

All ships are listed in *Italic;* all other entries are listed without the use of *Italics.*

Ship's names are often longer than the name commonly used as *Infanta Maria Teresa* is usually referred to as *Maria Teresa* When this occurs both names are indexed with the less commonly used name directing the reader to the common name, i.e. *Infanta Maria Teresa* will direct to *Maria Teresa.*

Names are also occasionally duplicated and when this occurs, the date of launching or other identifying characteristic is provided in the index or ships tables to aid in distinguishing them, i.e. *Orel* (hospital ship) or *Esmeralda* (Launched 1855)

One further note, occasionally a ship will bear the name of an outstanding person who participated in the history presented here. Admiral Condell for instance was honored by his country, Chile, by having a vessel named for him: *Admirante Condell.* Both the Admiral and ship are listed in the index and may be distinguished by the Italics used for the vessel. Incidentally, when Carlos Condell was serving his country he had the rank of captain and is so indexed. He was promoted to admiral and the ship bears that title.

Spelling is the common phonetic anglicized version of the ship's name particularly for the Russian and. Asian names. For lack of a better place to insert this, I will remind the reader that in traditional Asian names, the surname is placed first followed by the individual name (what would be termed the "first name" in western indices). Thus, Admiral Ting Juchang has a family name Ting and an individual name of Juchang. Here the index sites Ting, Admiral Juchang.

A

E

H

S

U

V

About the Author

Peter Stetson was born on November 4, 1947 and died on November 29, 2001. He attended Northeastern University in Boston, MA and went on to pursue his intense interest in naval history. This book is the ultimate result of that passion and pursuit; unfortunately published after his death.

About the Co-Author and Editor

Paul Janson is a graduate of Boston University's College of Engineering and of its School of Medicine. He is currently an emergency medicine physician practicing at Lawrence General Hospital in Lawrence, MA. He lives in Georgetown, MA with his wife, Mary, of 40 years and two daughters. As he approaches retirement from medical practice he has become engaged in other pursuits, writing being among them. He has written a children's picture book about the adoption of his daughters titled: *The Child In Our Hearts* and his first novel: *Mal Practice, a mystery of medicine and murder,* has been given a Finalist Award by the IPNE (Independent Publishers of New England) in the 2014 Book Contest, Genre Fiction category

Editor photo by Sarah McGrath

Other books by Paul Janson include

Mal Practice *a mystery of medicine and murder*. Pediatrician Joe Nelson is being sued in a malpractice case and getting divorced too. Thing get worse when he discovers his patient was murdered and he is next on the killer's list. He survives, and the sequel, **With a Little More Practice,** is now available.

Scratch *a young adult novel of magic and cats* A magical cat saves lives and changes lives with a magical scratch.

The Ice Cream War *a mystery of hot fudge and murder*. The story of two rival ice cream shops and the body found in one of them.

The Child In Our Hearts. A children's picture book about adoption, based on the belief that all children begin in the heart of their parents regardless of how they come to be in a family and regardless of what kind of family they become a part of. There are several versions of this book for different families, including two mothers and two fathers, and single parents. Some are available in Spanish as well. There is also a version for Assisted Reproductive Technology births.

www.ingramcontent.com/pod-product-compliance
Lightning Source LLC
Chambersburg PA
CBHW060542200326
41521CB00007B/449